MANUFACTURING SYSTEMS

FOUNDATIONS OF WORLD-CLASS PRACTICE

JOSEPH A. HEIM and W. DALE COMPTON, *Editors*

Committee on Foundations of Manufacturing
National Academy of Engineering

NATIONAL ACADEMY PRESS
Washington, D.C. 1992

NATIONAL ACADEMY PRESS • 2101 Constitution Ave., NW • Washington, DC 20418

NOTICE: The National Academy of Engineering was established in 1964, under the charter of the National Academy of Sciences, as a parallel organization of outstanding engineers. It is autonomous in its administration and in the selection of its members, sharing with the National Academy of Sciences the responsibility for advising the federal government. The National Academy of Engineering also sponsors engineering programs aimed at meeting national needs, encourages education and research, and recognizes the superior achievement of engineers. Dr. Robert M. White is president of the National Academy of Engineering.

This publication has been reviewed by a group other than the authors according to procedures approved by a National Academy of Engineering report review process.

Partial funding for this effort was provided by the Intel Foundation and the National Academy of Engineering Technology Agenda Program.

Library of Congress Cataloging-in-Publication Data

Manufacturing systems : foundations of world-class practice / Joseph A. Heim and W. Dale Compton, editors.
p. cm.
Includes bibliographical references and index.
ISBN 0-309-04588-6 (hard); ISBN 0-309-04678-5 (soft)
1. Manufactures—Management. 2. Industrial management. I. Heim, Joseph A. II. Compton, W. Dale.
HD9720.5.M38 1992
658.5—dc20 91-36171
CIP

Printed in the United States of America

Foreword

Manufacturing is a complex activity drawing upon many disciplines and technologies, reflecting management attitudes and philosophies, dependent upon organizational structures, influenced by the customers for products and by the suppliers of the many materials, machines, and equipment used to produce those products. Most efforts to develop a science of manufacturing have concentrated on understanding and improving the performance of unit operations and activities, in the belief that maximizing the effectiveness of the separate parts would result in an optimized system. Although our endeavors have provided a greater understanding of the fundamental phenomena underlying the individual components and an increased awareness of the details needed to direct and control them, we are beginning to realize that the complexity of the myriad relationships, interactions, and dependencies of the components and processes precludes such an approach to system optimization. It is now clear that ignoring the many interactions prevents good predictions of system performance and improved controls.

The study on which this report is based was an effort by the National Academy of Engineering to define principles underlying the complexity of manufacturing systems. Many of the findings of this study were presented at an NAE symposium entitled "Foundations of World-Class Manufacturing Systems" held in Washington, D.C., on June 19, 1991. I would like to thank W. Dale Compton, who chaired the symposium and the Committee on

Foundations of Manufacturing, and Joseph A. Heim, the principal staff officer for the project, for their efforts in organizing the symposium and in the publication of this volume. Also, on behalf of the National Academy of Engineering, I would like to thank the committee members (listed on page 254) and the authors who participated in the study and symposium. Special thanks are due to the following individuals in the NAE Program Office who worked on the publication and symposium: H. Dale Langford, Maribeth Keitz, Melvin J. Gipson, Vivienne T. Chin, and Mary J. Ball.

This project was carried out under the auspices of the NAE Program Office, directed by Bruce R. Guile. Partial funding for this effort was provided by the Intel Foundation and the National Academy of Engineering Technology Agenda Program.

ROBERT M. WHITE
President
National Academy of Engineering

Preface

Recognition of the complexity of the manufacturing system and the need to view it in its entirety was a principal outcome of the February 1987 National Academy of Engineering conference on "Design and Analysis of Integrated Manufacturing Systems: Status, Issues, and Opportunities." The conference identified a need for a more intensive search for the elements of a manufacturing systems discipline, based on the belief that a more effective characterization of manufacturing systems would benefit both practitioners and educators.

In response, the National Academy of Engineering assembled the Committee on Foundations of Manufacturing in early 1989 to address the disciplinary nature of manufacturing systems. The concept of manufacturing as a discipline implies the existence of "basic laws," a taxonomy, basic theories of design, and optimization procedures. This study has focused on the systems aspects of manufacturing and has been primarily concerned with those generic issues reflected in world-class manufacturing companies and is not intended to address operational solutions to today's manufacturing problems.

The Foundations of Manufacturing Committee met on numerous occasions during 1989 and 1990 to explore the various aspects of manufacturing systems, often meeting with manufacturing experts who are not members of the committee. Individual committee members, as well as the committee, also developed a variety of characterizations of manufacturing systems and subsystems and began work on developing or obtaining data to illustrate some of the hypothesized relationships. In July 1990 the committee hosted a workshop at the NAS/NAE study center in Woods Hole, Massachusetts, for several manufacturing executives, practitioners, and university educators. The exchange of ideas during the workshop lent further credence to the need to identify the core set of principles or bases on which manufacturing systems could be analyzed, designed, and managed. The enthusiastic reception for the concept and support for the effort to define and generalize these principles led to the development of several papers (included in this volume) that focus on many of the ideas presented during the workshop. The committee draws heavily on these papers in its arguments concerning the importance of these principles and their application in the manufacturing environment.

The committee designated these principles "foundations" of manufacturing because of their comprehensive applicability; they are generic, not specific to a particular industry or company; they are operational in that they lead to specific actions and show directions that should be taken; and their application should lead to improved system performance. These operating principles must be recognized, understood, and aggressively adopted by manufacturing organizations that aspire to world-class performance standards.

This report provides a framework that can be used by manufacturing executives and practitioners to improve their capability to predict the outcomes of product, process, and operating decisions and to assist them in analyzing, designing, and controlling their systems. For educators and those engaged in research, the report identifies opportunities for greater exploration and the discovery of additional foundations of manufacturing systems. Furthermore, it reminds us that intense interdisciplinary interactions of the many contemporary engineering, business, and social disciplines needed by modern manufacturing systems will contribute significantly to the global competitiveness of manufacturing in the United States.

W. DALE COMPTON, *Chairman*
Foundations of Manufacturing Committee
National Academy of Engineering

Contents

REPORT OF THE COMMITTEE ON FOUNDATIONS OF MANUFACTURING **1**

Executive Summary 3
1. Introduction 9
2. Overview 14
3. Management Practice 27
4. Measuring, Describing, and Predicting System Performance 43
5. Organizational Learning and Improving System Performance 61
6. Educational and Technological Challenges 78

GLOBALLY COMPETITIVE MANUFACTURING PRACTICES **83**

Involvement and Empowerment: The Modern Paradigm for Management Success 85
Nancy L. Badore

Implementation Projects: Decisions and Expenditures 93
H. Kent Bowen

Benchmarking 100
W. Dale Compton

Improving Quality Through the Concept of Learning Curves 107
W. Dale Compton, *Michelle D. Dunlap*, and *Joseph A. Heim*

Organizing Manufacturing Enterprises for Customer Satisfaction 116
Harry E. Cook
Customer Satisfaction 128
Harold E. Edmondson
The Interface Between Manufacturing Executives and Wall Street Visitors—Why Security Analysts Ask Some of the Questions That They Do 137
Philip A. Fisher
Taylorism and Professional Education 149
John E. Gibson
The Integrated Enterprise 158
William C. Hanson
Time as a Primary System Metric 166
Dan C. Krupka
Communication Barriers to Effective Manufacturing 173
James F. Lardner
Are There "Laws" of Manufacturing? 180
John D. C. Little
Taking Risks in Manufacturing 189
David B. Marsing
Constant Change, Constant Challenge 196
Joe H. Mize
Manufacturing Capacity Management Through Modeling and Simulation 204
A. Alan B. Pritsker
The Power of Simple Models in Manufacturing 215
James J. Solberg
Improving Manufacturing Competitiveness Through Strategic Analysis 224
G. Keith Turnbull, Eden S. Fisher, Eileen M. Peretic, John R. H. Black, Arnoldo R. Cruz, and *Maryalice Newborn*
Going to the Gemba 233
Albertus D. Welliver
Jazz: A Metaphor for High-Performance Teams 238
Richard C. Wilson

Consolidated Bibliography 245

Committee Membership 254

Biographies of Contributing Authors 256

Index 263

MANUFACTURING SYSTEMS

REPORT OF THE COMMITTEE ON FOUNDATIONS OF MANUFACTURING

W. Dale Compton, *Chairman*
H. Kent Bowen
Harry E. Cook
James F. Lardner
A. Alan B. Pritsker

Executive Summary

The objective of this study is to identify ways to make the system of manufacturing so efficient, so responsive, and so effective that it will make the organization of which it is a part the most competitive in the world. The committee concludes that this will be possible only through an enhanced understanding of the manufacturing system and a willingness to persist in a continuous examination of the conventional wisdom for managing and controlling that system. It is of prime importance that manufacturers constantly remind themselves that the adoption of a "system view" is critical to accomplishing the desired objectives.

This study argues that the modern manufacturing organization cannot be competitive if it continues to operate as a loosely coalesced group of independent elements whose identity depends on a discipline or a detailed job description. The study explores principles that have been demonstrated as generic to improving the effectiveness of manufacturing systems. The study draws heavily on the experience of U.S. manufacturers and a rapidly growing body of scholarly work linked closely to changing industrial practices. Particularly relevant research and publishing includes such topics as concurrent product and process engineering, total quality management, just-in-time manufacturing and distribution processes, quality function deployment, lean production, and incorporation and management of innovation.

With the assistance of many experts and practitioners who have participated in meetings and workshops and written papers as background material for this volume, the committee has identified a group of operating principles that must be recognized, understood, and adopted by manufacturing organizations that aspire to be "world class." Because of the universality of these principles, the committee has designated them as "foundations" of manufacturing. These foundations are generic in that they are not specific to a particular industry or company; they are universal in that they can be applied in a wide variety of circumstances; they are operational in that they lead to specific actions and show directions that should be taken.

It is the committee's strongly held conviction that the worldwide competitive environment will richly reward the manufacturers who adopt these principles while penalizing those who do not. Although success in implementing the foundations depends on many things, the committee emphasizes that they represent a system of actions that cannot be embraced piecemeal. The foundations are as interrelated and as overlapping as are the elements of the manufacturing system they are intended to improve. The foundations must be viewed as a system of action-oriented principles whose collective application can produce important improvements in the manufacturing enterprise.

Goals and Objectives

What should be a manufacturer's goal if it is to compete successfully in the global marketplace? This goal is often referred to as being a "world-class" manufacturer, a term used to convey the sense of excelling. The Japanese describe it as striving to be the best-of-the-best.

> FOUNDATION: World-class manufacturers have established as an operating goal that they will be world class. They assess their performance by benchmarking themselves against their competition and against other world-class operational functions, even in other industries. They use this information to establish organizational goals and objectives, which they communicate to all members of the enterprise, and they continuously measure and assess the performance of the system against these objectives and regularly assess the appropriateness of the objectives to attaining world-class status.

The Customer

A manufacturing organization serves a variety of customers. In addition to the customers who expect to purchase high-quality products and services, the owners or stockholders may also be thought of as customers in that they expect a reasonable return on the investment that they have made in the company. The employees are customers in that they expect an employer to recognize their contribution to the success of the company and to provide them with a reasonable reward for their efforts. These are the stakeholders in the organization in that each has made a personal commitment to its success. The stakeholders have special expectations and needs that must be met.

> FOUNDATION: World-class manufacturers instill and constantly reinforce within the organization the principle that the system and everyone in it must know their customers and must seek to satisfy the needs and wants of customers and other stakeholders.

The Organization

The complexity of the manufacturing system arises from many directions: the interdependence of the elements of the system, the influence of external forces on it, the impact that it can have on its environment, and the lack of predictability in the consequences of actions.

> FOUNDATION: A world-class manufacturer integrates all elements of the manufacturing system to satisfy the needs and wants of its customers in a timely and effective manner. It eliminates organizational barriers to permit improved communication and to provide high-quality products and services.

The Employee

Creating a world-class manufacturing organization begins with recognition that the most important asset of the organization is its employees. When properly challenged, informed, integrated, and empowered, the employees can be a powerful force in achieving the goals and objectives of the organization.

> FOUNDATION: Employee involvement and empowerment are recognized by world-class manufacturers as critical to achieving continuous improvement in all elements of the manufacturing system. Management's opportunity to ensure the continuity of organizational development and renewal comes primarily through the involvement of the employee.

The Supplier or Vendor

It is essential that the barriers that have existed between supplier and purchaser be attacked as actively as are the barriers between the elements in a manufacturing organization. The sharing of goals, the exchange of information, the interchange of people, and the making of long-term commitments are some of the ways in which these barriers are being overcome.

> FOUNDATION: A world-class manufacturer encourages and motivates its suppliers and vendors to become coequals with the other elements of the manufacturing system. This demands a commitment and an expenditure of effort by all elements of the system to ensure their proper integration.

The Management Task

Imaginative, creative leadership at every level of an organization is critical to building on the foundations of world-class manufacturing systems. Management creates the culture within which the organization functions. Management must exhibit the concern for the health and well-being of the organization's human resources. Management must insist that the organization look beyond its borders to interact with its customers, its suppliers, and the educational systems that are training its present and future employees. It is a challenge to the organization to find the proper management for the circumstances in which it finds itself.

> FOUNDATION: Management is responsible for a manufacturing organization's becoming world-class and for creating a corporate culture committed to the customer, to employee involvement and empowerment, and to the objective of achieving continuous improvement. A personal commitment and involvement by management is critical to success.

Metrics

Performance evaluation is a process applied throughout the manufacturing organization to measure the effectiveness in achieving its goals. Because of the variety, complexity, and interdependencies found in the collection of unit processes and subsystems that define the manufacturing system, appropriate means are needed to describe and quantify rigorously the performance of each activity.

> FOUNDATION: World-class manufacturers recognize the importance of metrics in helping to define the goals and performance expectations for the organization. They adopt or develop appropriate metrics to interpret and describe quantitatively the criteria used to measure the effectiveness of the manufacturing system and its many interrelated components.

Describing and Understanding

It is difficult to conceive of improving the current status of the system without first having a clear description of its status and character. This requires identifying the interrelationship and theoretical limits of the operational variables. It demands that important system parameters be identified and measured.

> FOUNDATION: World-class manufacturers seek to describe and understand the interdependency of the many elements of the manufacturing system, to discover new relationships, to explore the consequences of alternative decisions, and to communicate unambiguously within the manufacturing organization and with its customers and suppliers. Models are an important tool to accomplish this goal.

Experimentation and Learning

Organizational learning is a broad-based strategy for capturing and making available to members of the organization information and knowledge that enable them to benefit from experimentation and the experience of others. Too often in manufacturing, sources of information become scattered and isolated and individual learning experiences are not automatically converted to organizational memory and made available for all members to draw and build upon. The rate at which an organization improves its performance as a result of learning is perhaps one of the principal determinants of whether it can become best-of-the-best.

> FOUNDATION: World-class manufacturers recognize that stimulating and accommodating continuous change forces them to experiment and assess outcomes. They translate the knowledge acquired in this way into a framework, such as a model, that leads to improved operational decision making while incorporating the learning process into their fundamental operating philosophy.

Technology

U.S.-based manufacturers have often adopted the view that technological prowess is a viable means of compensating for other shortcomings. It is the committee's strong conviction that a manufacturer can make the best use of technology only after it has embraced and is practicing the foundations described above. Only then can technology become a powerful force in achieving a competitive advantage.

For management, selection of the proper technologies from among technological opportunities is becoming a complex challenge that may be different for each manufacturer and for individual facilities. Each manufacturer must develop a strategy that encourages the search for the best and most important technologies, develops a procedure for effectively analyzing technological opportunities, creates or acquires the expertise needed to implement those technologies, and commits the necessary financial and human resources to introduce the new developments when they become available.

> FOUNDATION: World-class manufacturers view technology as a strategic tool for achieving world-class competitiveness by all elements of the manufacturing organization. High priority is placed on the discovery, development, and timely implementation of the most relevant technology and the identification and support of people who can communicate and implement the results of research.

Implementing the Foundations

The implications of the competitive environment that has evolved over the past 20 years are profound. Just as no single element in the manufacturing system can ensure that an enterprise will be successful, so can no single sector of the national infrastructure ensure that the industrial sector will be competitive. A commitment to renewal of the U.S. manufacturing sector is essential. A willingness to learn from each other is critical. No one can afford to take the risk of waiting for others to show the way. All manufacturers must embrace the doctrine that continuous improvement demands their immediate and unrelenting attention. U.S. manufacturers cannot allow their competitors to set the standards by which success will be achieved and to be the leaders in meeting those standards. In addition, the United States must establish as a national goal a strategy that encourages and supports the adoption of the foundations of world-class manufacturing systems.

1

Introduction

> The world hates change, yet it is the only thing that has brought progress.
>
> —Charles F. Kettering (1876–1958)

The evolutionary path taken by modern civilization has been closely paralleled—one might even say that civilization has been propelled along that path—by the ever-increasing improvement in the efficiency with which materials have been converted into forms that enhance the standard of living for the human user. In early times, manufacturing was carried out by an individual artisan who created a product for a specific user. The attributes of the product were tailored to meet the needs of the user. The modern manufacturing system—a complex arrangement that engineers, manufactures, and markets products for a diverse populace whose wants and needs are subject to frequent change—bears little similarity to its early predecessor.

This transformation of the manufacturing function into the complex technical, social, and economic organization it is today has been made possible by the discovery and improvement of many methods and processes. Some of these have revolutionized the technologies employed in the transformation of materials; some have made possible the creation of sophisticated organizations for the design, production, and marketing of products and services; and some have made it possible to create new and unique materials that have expanded design alternatives and formed the basis for entirely new products, processes, and industries.

These developments have their origins in many activities, some internal to manufacturing and some well removed from it. It is not an exaggeration, however, to claim that some of the most profound changes that have affected the manufacturing enterprise can be traced to major developments that resulted from the complex interactions of society, technology, and the economy. Three of the most important of these are widespread and inexpensive transportation systems, communication systems that provide real-time interaction between people in almost any part of the world, and computers that assist in the design, control, and analysis of complex activities.

Given the ability to ship most finished goods quickly and inexpensively almost anyplace in the world, companies no longer must locate their factories in or near the markets they serve. Neither do manufacturers require an extensive national industrial infrastructure to support local or regional factories with low levels of manufacturing integration. Modern communication technology has made it possible to manage manufacturing operations located around the world. A highly dispersed network of facilities and suppliers can now be created irrespective of their location. For example, the local availability of raw materials is not a determinant or predictor of successful manufacturing capabilities. Information concerning technological developments flows virtually unimpeded across national boundaries. Finally, the availability of modern computer technology has made it possible to analyze, design, and control the complex systems that characterize modern manufacturing operations. These three technical developments have had a profound effect on both the organization of the manufacturing enterprise and the strategies that manufacturing enterprises follow to survive.

A consequence of these new capabilities is that manufacturers can now establish an effective presence in almost any market that will accept their products. Many manufacturers have been aggressive in entering markets far from their domestic bases, particularly here in the United States. The extent of the challenge that U.S.-based manufacturing faces is reflected in the recent estimate that 70 percent of U.S. manufacturing output currently faces direct foreign competition (see National Research Council, The Internationalization of U.S. Manufacturing, 1990).

The increased competition that U.S.-based manufacturers experienced included other dramatic changes, all of which required a change in actions and attitudes if an enterprise hoped to remain competitive. Customers became unwilling to accept low-quality products. They expected and sought manufacturers who could offer them new products in a timely manner. Markets fragmented. Product life cycles shrank. New arrangements between vendor and purchaser were needed. Technological capability was expanding rapidly with the progress in firms and laboratories around the world. The old way of doing business—of treating markets and competition as local—was no longer adequate. A manufacturer who was unwilling or incapable of adjusting to the new competitive environment was unlikely to survive. The response by U.S.-based manufacturers to these challenges has been broad and, on balance, largely encouraging. Product quality has been improved. Management practices are being changed to make better use of existing human resources. Manufacturing efficiencies are being increased through more effective selection and management of physical and human resources. Responsiveness to changing market conditions and customer needs has been enhanced. Industry coalitions are being formed to explore the opportunities and early ramifications of emerging technologies. Joint university-industry programs provide new educational experience to students while focusing university research efforts on problems that are relevant to industry (see Schonberger, 1986, 1987, and Stewart, 1991). Industry consortia, such as SEMATECH and the National Center for Manufacturing Sciences (NCMS), have been organized to fill the gaps in critical technical areas.

Although these responses are encouraging, the competitive environment is constantly changing. Manufacturers around the world continue to make important product and process innovations and improvements. To survive in this competitive environment, an aggressive and effective program of continuing improvement involving all parts of the manufacturing enterprise must be established. To give focus to such a program, manufacturers must continuously scan the world to determine what constitutes the "best existing manufacturing practice" and who is the best manufacturer in the world (see Edmondson and Wheelwright, 1989). It is only by identifying these best practices that direction for continuing improvement can be provided. Only then will a manufacturer be able to approach "world-class" status and be able to survive in the global competitive environment. This does not imply that a manufacturer should copy and implement the best practices of others. Instead, manufacturers should learn from each other and seek to incorporate best practices as appropriate to their particular organizations.

While it is a relatively straightforward process to identify the goals that a continuing improvement program should achieve, it is far more difficult to determine the steps needed to achieve those goals. A large part of the difficulty lies in the fact that a manufacturing company is a very complex system comprising a large number of complex but interdependent subsystems. Only a limited amount of work has been done to understand the problems of managing the totality of the manufacturing system and to identify the degree and nature of the interrelationships of its subsystems.

This study focuses on the manufacturing system as an entity. It explores principles that have been demonstrated as generic to improving the effectiveness of manufacturing systems. It draws on the experience of many manufacturing experts and practitioners who participated in meetings and workshops and prepared papers for this volume. With their assistance, and acknowledging the growing body of literature in the fields of manufac-

A Simple Game Played Very Quickly

Economic warfare is the metaphor often used to describe the environment confronting globally competitive manufacturers. I would prefer an alternative metaphor to warfare: The game we're going to play in the 21st century is some combination of world class chess and what the world knows as football (and Americans know as soccer). American football has three characteristics that soccer doesn't have: huddles, timeouts, and unlimited substitution. Soccer has no hurdles, no timeouts, and very limited substitutions. We're going to play a much faster game than we've been playing in the past. World class chess is a very simple game. The one who can think farthest ahead will win. Designing a strategy for the future doesn't play to an American strength.

SOURCE: Thurow (1992).

turing technology, quality, management practices, and information systems, the committee has identified a group of operating principles that must be recognized, understood, and adopted by manufacturing organizations that aspire to be "world class." Because of the universality of these principles, the committee has designated them as "foundations" of manufacturing. These foundations are generic in that they are not specific to a particular industry or company; they are universal in that they can be applied in a wide variety of circumstances; they are operational in that they lead to specific actions and directions that should be taken.

An elucidation of all the key foundations of manufacturing is well beyond the scope of the current study. No single study can expect to address all of the foundations for a field as diverse as manufacturing. What the committee seeks here is a framework for, and a description of, some of the principles that should be included in the foundations of manufacturing. Confirmation and enlargement of the views presented here are recognized as a long-term process worthy of much further intellectual study and pragmatic research.

It is hoped that although this book does not provide definitive answers, it will provide some important insights into that broad topic described as the System of Manufacturing and that it will serve as an encouragement to readers to develop further understanding of the challenge of creating an effective, world-class manufacturing organization. With greater understanding, it will be easier to set goals, to develop plans that will command consensus, and to choose directions that will achieve continuing improvement of the system.

This study is concerned with actions; actions that an organization must take if it is to become a world-class manufacturer. Resources—human, financial, managerial, and technical—must be effectively organized and used. Metrics must be established by which progress in continuous improvement can be measured. The organization must gain enough understanding of the manufacturing system and enough confidence in that knowledge to be willing to experiment and to learn. Ultimately, it is necessary to be able to predict the manufacturing system's response to changes. Technology must become a key factor in achieving and sustaining a competitive advantage. Organizational objectives, both short- and long-term, must be understood and accepted at every level without becoming distorted by functional bias. Finally, leaders must be effective and have the courage to lead. Not only must an organization meet all of these challenges if it is to become the "best-of-the-best," but the constantly changing environment also requires that an organization continue to change and improve to retain that capability.

2

Overview

The factory is a human phenomenon. Every step from conception to eventual destruction is for, by, and because of people.

—G. Nadler and G. Robinson, 1983,
in *Design of the Automated Factory: More Than Robots.*

The study of the manufacturing enterprise has long interested researchers. Management and organizational aspects of manufacturing have been examined extensively. A "science" of manufacturing has focused largely on its separate components—such as material handling, material transformation, plant layout, and the data and information systems—that support the various manufacturing functions. The combination of these many efforts has been of great value in that they have contributed to a greater understanding of the details of the manufacturing process and have broadened the understanding of the fundamental phenomena that control the components of manufacturing.

What we have come to realize, however, is that an understanding of the separate unit operations of manufacturing, no matter how complete, is not sufficient. The manufacturing system is much greater than the production facilities or the transformation processes used there. It includes all the functions and activities that relate to the conception, design, making, selling, maintaining, and servicing of the product. Manufacturers must constantly remind themselves that maintaining this "system view" is critical to understanding the totality of these functions and the interrelationships among them—in short, the total enterprise (see Merchant, 1988). It is also important to recognize that although the need for systems emphasis is acute in many instances, it should be balanced with considerations for the unit operations.

In the absence of an understanding of the totality of the manufacturing system, operational paradigms have often evolved from beliefs or rules of thumb that derive from personal experience or individual interpretation of empirical data gathered from day-to-day operations in uncontrolled environments. This so-called know-how varies widely in extent and validity from company to company and from industry to industry. It is frequently situation dependent and, therefore, often impossible to generalize or to apply to new situations.

This lack of emphasis on system issues is not the result of a lack of appreciation for the importance of the problem. Rather, anyone attempting to address these system issues is immediately confronted by the overwhelming complexity of the problem. Manufacturing systems are a complicated combination of physical systems and human workers and managers. The tools for treating large, complex systems are limited (see Hatvany, 1983, and Senge, 1990). Data on the performance of manufacturing systems are often fragmentary and incomplete, and even where the data are excellent, competitive pressures prevent the data from being disseminated and made available for research. Moreover, metrics used to evaluate the performance of the manufacturing enterprise seldom address system performance.

In this overview we define some key terms, indicate how the foundations might be used, and identify the likely audience for this volume. A detailed discussion of each of the foundations follows in subsequent chapters.

THE MANUFACTURING SYSTEM

The manufacturing system can mean many things, depending on the viewpoint taken. Figure 1 presents the committee's view. Operations placed at the center of the enterprise, overlapping and interacting with administration and management, the product and process engineering activities, the applied sciences, and the marketing, sales, and service activities. Overlapping and interacting with all of these functions are the customers for the products or services; the vendors and suppliers that provide materials, components, and services to the enterprise; the community in which the enterprise exists; and the government that establishes regulations, rules, and opportunities for the enterprise.

Figure 1 is intended to illustrate the interrelationships that exist in the manufacturing system. Although particular technologies are not identified

Developing a Science of Manufacturing

The problem is not, . . . simply one of applying existing technology in a systematic way. The problem is to develop a genuine science of manufacturing. This need is not well understood, perhaps because of the common misconception that the natural progression of things is from science to engineering to application, or from basic science to applied science to development. In fact, history is full of examples of technology and engineering getting ahead of science, followed by the creation of the science base, which in turn allows refinements in the technology.

The classic example is the steam engine and thermodynamics. But there are many other and more timely examples of the same phenomenon:

- A workable technology of photography was developed by artists and craftsmen decades before physicists and chemists understood the principles involved. Modern photography, however, rests firmly on a scientific base.
- Communication theory and computer science grew out of engineering approaches and real-life experiences of noise in communications channels and computer design.

In manufacturing we have a technology analogous to Watt's steam engine and to early photography. We know enough to do useful things, but there is a real limit to how far we can go. We have only limited knowledge of efficient dynamic scheduling algorithms. We don't know how to automatically cope with machine breakdown. In short, we have no theory to guide us in the efficient design and construction of an optimized system.

We do, however, know enough to ask good questions about the principles involved—to begin to develop the science. But that science is in its early infancy. We must develop the science of design and manufacturing from the ground up, basing it on models and physical laws just as we would any other science. We must codify our experiences to be able to develop generalized insights and approaches.

The new theory that we need will be truly interdisciplinary and will draw on several engineering disciplines, as well as on computer, information, and materials science. The integration of these disciplines will require new institutional arrangements in our universities.

SOURCE: Bloch (1985).

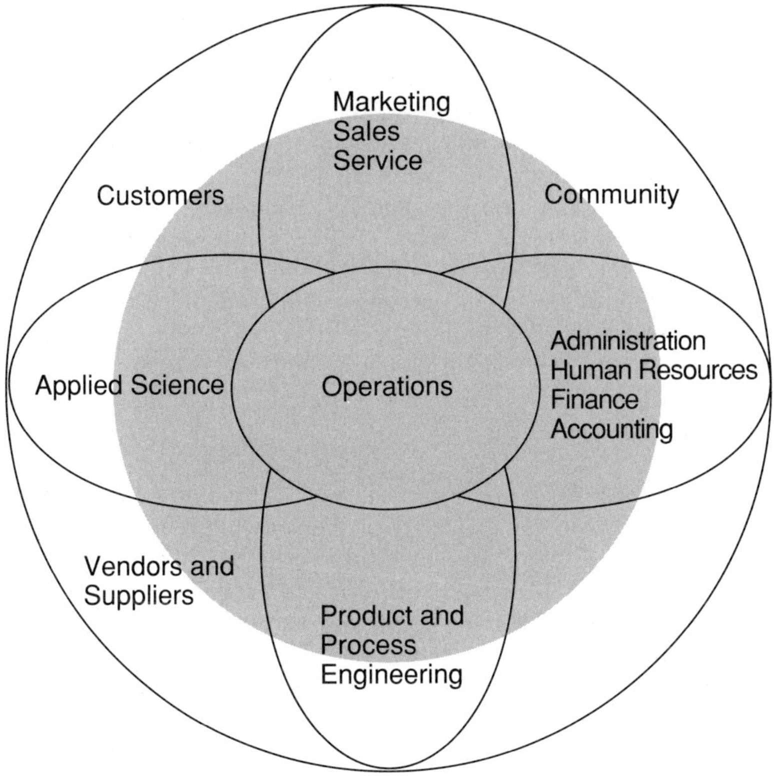

FIGURE 1 The integrated manufacturing enterprise. Overlapping functions, disciplines, and activities.

in this description of the manufacturing system, it must be understood that they establish or enhance many of the capabilities of the various functions contained in the system. As noted, the applied sciences provide the technical base for many of the areas. The material transformation processes, sometimes referred to as unit processes, are the means used to convert materials into components and subsystems. Computer-based systems provide the tools to enhance the capability and performance of the design, planning, scheduling, control, and sales of the products. The product and process engineering, the unit manufacturing operations, the marketing, sales, and services, the vendors and suppliers, and the management and administration each benefit from these systems and their capabilities to describe system performance. Describing the total enterprise in this way draws attention to the fact that no single unit operation or function can exist in isolation from all other components of the system. It is the realization of the interdependencies among these many components of the system that has created the impetus for "simultaneous engineering" or "concurrent engineering." A successful product realization process demands recognition of these interdependencies and the overlapping of interests among unit operations. (further discussions of simultaneous engineering and the product realization process can be found in National Research Council, The Competitive Edge: Research Priorities for U.S. Manufacturing, 1991a, and Improving Engineering Design, 1991b). Figure 1 also suggests how different viewpoints, values, and objectives for the system can develop, depending on the discipline or functional group in which individuals work. People working in applied science, finance, marketing, service, or engineering may have very different views about their role and where it fits in the system. They may also have very different perspectives about the system than do the production people on the factory floor.

An important conclusion to be drawn from this diagram is that if a manufacturing enterprise is to succeed, there can be no basic difference in viewpoints, values, and goals among its constituent groups. It is clear that the areas of responsibility are not neatly separated from one another but overlap to an important degree; financial and accounting systems, for example, have a major impact on operations and engineering. Perceived or artificially created boundaries between organizational units, such as those between marketing and engineering, production and purchasing, production control and marketing, or employees and management, both restrict and complicate communication and cooperation. The performance of the system suffers. The challenge to management is to find ways to take advantage of the strengths of the various unit operations and functional groups while discouraging any tendencies to work at cross purposes or toward conflicting goals (see Dertouzos et al., 1989). Achieving true involvement among the various activities requires, of course, more than simply reducing the barriers between groups.

The manufacturing system is a complex organism. It receives inputs from the outside world (product and service concepts, orders, materials, and energy), uses a set of resources to respond to those inputs, transforms materials or components into a form that is needed or desired by a customer, and operates within constraints that are determined by physical, financial, human, and political limitations. By further challenging the traditional view of manufacturing as a collection of activities and functions, and recognizing the overlapping and interconnected disciplines involved, we realize the need for a dynamic and integrated concept of manufacturing systems. Given the dynamism of the current manufacturing environment, the manufacturing system might be viewed as a collection of transformation subsystems that must be properly integrated. These might include the materials transformation subsystem; the customer needs transformation subsystem; the knowledge, learning, and improvement subsystem; and the organizational renewal subsystem. Focusing on the integrated system, as opposed to the individual functional parts that make up the system, is critical to understanding the key relationships and interactions in the overall performance of the enterprise.

Krupka observes (in this volume, p. 166) that "it is necessary to recognize that manufacturing operations—the activities that take place within the walls of a factory—can no longer be treated as the system to be optimized." Instead, it is necessary to consider manufacturing as only one of several systems:

> Customers' orders for products are conveyed by an ordering system to the manufacturing system, whose output then flows through a distribution system to the customer. Rapid and flexible response requires that materials and parts flow quickly into manufacturing; that requires a short and predictable interval for the material provisioning system. In addition to ensuring high performance for these systems, a successful firm must be capable of rapidly translating its designs into manufactured products. Hence, we need a well-crafted and rapid product introduction system.

Gibson's view (in this volume, p. 149) is that

> Objective observers are becoming increasingly aware of the need to consider the manufacturing process as a whole rather than as an object for piecewise suboptimization. This holistic viewpoint must include manufacturers' relations with subcontractors and suppliers as well as customers. The manufacturing system certainly must include the interrelationship of the physical manufacturing environment, manufacturing management, and the worker.

Space does not permit addressing in detail all unit operations or actions that are critical to the successful operation of the manufacturing system. While not wishing to minimize the importance of costs and profits in operating a competitive business, the committee addressed these objectives only indirectly. The committee also discussed only briefly the product realization process, although it is certain that the manufacturing activity must be involved in this in an important way. Moreover, the challenges posed by processes used to transform the state of materials or the technologies that underlie them are beyond the scope of this report.

It is important to recognize that among the multiple inputs to the manufacturing system, only some are predictable. The continuing challenge to manufacturers concerned with the competitive status of their industry is to encourage the exploration of the unpredictable inputs and their associated responses, to measure, to model, and to search for a level of understanding that will enhance their capability to optimize performance. Only by so doing is it possible to achieve that level of performance associated with world-class status. Over time the real differentiators are organizational learning and the requisite change and improvements.

MANUFACTURING FOUNDATIONS

Despite the many possible definitions and views of the manufacturing system, the committee's objective is clear. It is to find ways to make the system of manufacturing so efficient, so responsive, and so effective that it will make the entity within which it is embedded the most competitive in the world. This will be possible only through an enhanced understanding of the manufacturing system and a willingness to persist in a continuous examination of the conventional wisdom for managing and controlling that system.

In examining the actions and procedures that the most successful manufacturers have taken as they have evolved to world-class status, the committee notes that many of them have adopted common approaches. This suggests that manufacturers who aspire to world-class capabilities should understand and follow, to the extent possible, the successful approaches of others and learn to make their own improvements. It is to this end that this study was undertaken. The committee took as its task identification of the corpus of operating principles, which it has chosen to call the foundations of manufacturing, that are being used by world-class manufacturers. A number of experts have assisted the committee in identifying the rules, laws, or principles of practice that are applicable to all enterprises. The committee draws heavily upon the content of the papers prepared by these experts in its arguments concerning the importance of these issues and how they can be used daily in the manufacturing environment.

The foundations for a field of knowledge provide the basic principles, or theories, for that field. Foundations consist of fundamental truths, rules, laws, doctrines, or motivating forces on which other, more specific operating principles can be based. While the foundations need not always be quantitative, they must provide guidance in decision making and in operations. They must be action oriented, and their application should be expected to lead to improved performance. In the committee's view, the "foundations of manufacturing" should be universal to manufacturing industries—at least to companies in the same industry—and they should be culture free.

Examples of foundations can be found in many fields of engineering. The laws of thermodynamics are used to determine the theoretical limits of efficiency of various heat cycles. Maxwell's equations and quantum mechanics provide the electronic designer with the structure within which to understand and predict the operation of solid-state electronic components and systems. In the design of chemical reactors, the various laws describing fluid flow and mixing have led to the development of certain "scaling laws" that assist the designer in moving from a laboratory scale model to commercial-size systems. Linear and nonlinear mechanics form the basis for understanding the behavior of materials under load. Viscosity, boundary layer phenomena, and molecular surface phenomena are important elements of the foundations of lubrication.

The foundations of manufacturing differ in important respects from those just described. For the manufacturing system, one is dealing with a complex combination of disciplines and technology, management attitudes and philosophies, organizational issues, and the influences of an environment that includes the customers for the product that is being produced. In dealing with this complexity, the committee has constructed a framework that allows the foundations to be grouped under three topics:

- Foundations that relate to management philosophy and management practice (Chapter 3).
- Foundations that relate to the methods used to describe and predict the performance of systems (Chapter 4).
- Foundations that relate to organizational learning and to improving the performance of systems through technology (Chapter 5).

The foundations related to management explicitly recognize that actions, decisions, and policies advocated and implemented by all levels of management are critical determinants of the success of an enterprise. Included in this grouping is the critical operational philosophy that emphasizes the importance of continuous improvement of all operations in the enterprise and the importance of employee involvement in achieving this form of improvement. There is the role of employee empowerment in achieving the timely solution to problems. There are the interactions that the manufacturing enterprise must have with other activities in the company, with their suppliers, and with the customer. There is the importance or organizational structure, communications, and goal setting. While these elements of the foundations of manufacturing are not quantitative in the usual sense, it is abundantly clear that world-class manufacturers have generally recognized and are applying these management practices and that these practices have contributed critically to their success.

It is difficult to conceive of improving the current status of the system without first having a clear description of its status and character. This requires identifying the interrelationship and theoretical limits of the operational variables. It demands that important system parameters be identified and measured. Identifying cause-and-effect relationships that help predict the consequences of actions provides a basis for developing general tools and procedures that will allow the practitioner to extrapolate beyond current operating experience and to anticipate more accurately how a future system

The Principles of Scientific Management

It is true that whenever intelligent and educated men find that the responsibility for making progress in any of the mechanic arts rests with them, instead of upon the workmen who are actually laboring at the trade, that they almost invariably start on the road which leads to the development of a science where, in the past, has existed mere traditional or rule-of-thumb knowledge. When men, whose education has given them the habit of generalizing and everywhere looking for laws, find themselves confronted with a multitude of problems, such as exist in every trade and which have a general similarity one to another, it is inevitable that they should try to gather these problems into certain logical groups, and then search for some general laws or rules to guide them in their solution. As has been pointed out, however, the underlying principles of the management of "initiative and incentive," that is, the underlying philosophy of this management, necessarily leaves the solution of all of these problems in the hands of each individual workman, while the philosophy of scientific management places their solution in the hands of the management.

SOURCE: Taylor (1934).

may respond or perform. The extent to which modeling, simulation, and analysis can be developed to provide these capabilities is an important element of the foundations of manufacturing. Although some of these quantities could be explored through experiments in the laboratory, the committee recognizes that some may need to be validated by techniques similar to those employed in microeconomics, social science, and cultural anthropology.

The objective of maintaining and achieving enhanced system performance requires an environment in which an organization can learn and benefit from its past experiences. As operating practice becomes more efficient through the application of the foundations of manufacturing, technology will become a more critical element in maintaining the status of the world-class competitor. The arrangements for acquiring, developing, and introducing new technology will become increasingly important as U.S. manufacturers continue to develop their abilities to compete in the world marketplace.

The elements of the framework just presented are purposefully ordered, reflecting the committee's deeply held belief that world-class competitiveness can be achieved only by properly applying all of these foundations, starting with management and progressing through the foundations related to metrics and technology. Unless the foundations of management have been put in place, the remaining foundations are not likely to be of lasting benefit to the enterprise. While rules and laws—combined with the continual measurement of important operating parameters—provide the capabilities to set goals and measure progress, these are unlikely to have the desired effect unless the management issues have been addressed. Enhancing an organization's ability to learn from experience is critical to its success, but the value of this learning will depend on how well the enterprise is managed and how thoroughly it understands its current operations. While technology may well become the ultimate tool for achieving a competitive advantage, the success that an enterprise has in using it may depend on how well it adopts and integrates the other foundations.

THE BENEFITS OF FOUNDATIONS OF MANUFACTURING

What are the potential benefits of foundations for manufacturing? As noted above, these foundations should be viewed as operational guidelines—principles that can be applied in a wide variety of circumstances by those who wish to be a part of an enterprise whose goal is to be a world-class manufacturer. They represent criteria by which actions can be judged, goals and objectives established, and progress measured. In this regard, the following advantages appear to be realizable by applying the foundations.

First, a foundation provides a body of knowledge—a basis for understanding—that industrial and manufacturing executives could use to improve their ability to predict the outcome of specific product, process, and operating decisions. An immediate benefit should be the development of better generic tools for analyzing, designing, and controlling systems. One might hope, for example, that it would become common practice to explore thoroughly the impact of product complexity on the efficiency of the manufacturing operation instead of focusing only on the impact that additional product offerings will have on the marketing and sales activities.

Second, an understanding of the elements of a foundation should indicate some of the opportunities for more meaningful interdisciplinary interactions, for example, among scientists, engineers, production managers, and those who are associated with sales and marketing. Research programs and applications could share a common vocabulary, report on empirical measurements or experiments that test or validate new principles, and identify future research issues. The successful implementation of simultaneous engineering, for instance, critically depends on a common understanding of many of these interdisciplinary issues.

Third, a foundation can help guide the experimentation and learning process that is important to achieving future improvement. In addition, it can help focus the exploration and use of technology to improve a company's competitive position in the world marketplace.

A foundation may be in a primitive state, such as a collection of empirical observations, that relates variables or outcomes and assists the manufacturing leader with actions and the manufacturing researcher with a context for discovery. However, it must be recognized that an enterprise derives no great advantage from the identification and understanding of a foundation of manufacturing unless it recognizes the strategic importance of manufacturing. Turnbull and coauthors (in this volume), "place a special emphasis on strategic analysis in manufacturing." They observe that to "harvest the strategic implications of our manufacturing analysis, we recognize that we must look at a system that includes demand, competition/supply, and customer satisfaction." Thus, effective use of the foundations demands an organizational environment that encourages inclusion of manufacturing as a necessary strategic tool in becoming a world-class competitive force.

Some of the foundations that will be discussed below may be viewed by the reader as simply expressing common sense or good practice. The committee applauds those companies that have already adopted these foundations and encourages them to convey their experiences to others. The committee encourages those companies that are beginning to move toward achieving world-class performance to persist and to strive continuously to create an environment that fosters the involvement of all members of the organization in this important undertaking. There is ample evidence that the foundations will be of critical importance in meeting company objectives. Those companies that have not yet recognized the importance of these actions are urged to assess carefully their position relative to their world competitors. The committee strongly encourages these companies to cast aside previous attitudes and procedures and to embrace the foundations that follow.

THE AUDIENCE

The audience has, in a sense, been identified in the pages above. The committee believes that an understanding of, and appreciation for, these foundations will assist the corporate executive in asking critical questions of the organization and in arriving at the proper decisions. The executive responsible for manufacturing and the supervisors on the shop floor must understand the tools available to them for assessing alternatives and to be aware of the importance of organizational learning and the need for experimentation. In their efforts to optimize the individual operations, they must appreciate the interdependencies among the individual parts of the manufacturing system and the impact that changes in one part can have on other parts. Employees on the shop floor will benefit from an improved understanding of how their limited area of responsibility relates to the total system. Engineers and technologists, in their efforts to improve existing systems and to develop new capabilities, must appreciate how technology can help the system achieve world-class status.

Although this report tends to focus on the manufacturer, the committee

Competitive Manufacturing in the Next Century

How can the United States compete with countries that have abundant low-cost labor and are also aggressively developing and acquiring advanced technology?

The first requirement is that we accept that changes in manufacturing technology are inevitable. Instead of resisting these changes as we have tended to do in the past, we must find ways to take advantage of them. In terms of the directions for research, this means that we must investigate those technologies that can operate effectively in a changing environment. In more human terms, it means that we must emphasize the kind of education that prepares people for changing roles.

Second, we must understand the necessity of relying comparatively less on experience and more on sound theory. The ability to apply trial-and-error learning to tune the performance of manufacturing systems becomes almost useless in an environment in which changes occur faster than the lessons can be learned. There is now a greater need for formal predictive methodology based on understanding of cause and effect. Of course, a good deal of such methodology already exists, but the practices of industry tend to place greater reliance on experience-based knowledge than on theory-based knowledge. This difference is due in part to the failure of practitioners to familiarize themselves with the analytical tools that are available. In part it is due to a failure of the research community to develop the kinds of tools that are needed and to put them into a usable form. . . .

Another extremely important guiding principle for research is that we must generate reusable results having broad applicability. The best examples of advanced manufacturing systems that have been commercially developed, are tuned to the specific set of conditions in a single plant. Although these systems may provide great benefits to the companies who own them, there are few transferable benefits to the next company wanting to do something similar or even to the same company in another plant. In effect, each new system development project starts over from the beginning. If we are to have the kind of impact we desire on the whole of discrete manufacturing practice, we must find generic solutions that can be applied in many circumstances.

SOURCE: James J. Solberg, pp. 4-5 in Compton (1988).

believes that the educational community and the research community should also find these matters to be of interest. The implications of these foundations to the educational system, both in terms of content and organizational approach, are critical to the long-term health of manufacturing systems. The success of manufacturing is increasingly dependent on the availability and capability of well-trained people (see National Academy of Engineering, Education for the Manufacturing World of the Future, 1985). The communication of these foundations to the next generation of manufacturers, whether they will be managing the system, directing one of the unit operations, or developing new tools and technology, will increase the likelihood that the enterprise will achieve its objective. By focusing initially on the basic elements of manufacturing, instruction in manufacturing could build on verified principles, and the subsequent research could more easily be concentrated on broad system issues as distinct from detailed topics that have more limited applicability. It might be expected that this would lead the various university departments that focus on manufacturing and management of technology to offer a more consistent set of core courses and would result in a more systematic exploration of system-oriented topics by the students (further discussion in National Research Council, The Competitive Edge: Research Priorities for U.S. Manufacturing, 1991a).

For the research community, the committee wishes to encourage further study and examination of these topics. This report is offered as one step in encouraging others to enlarge upon these concepts and to critique the approaches suggested here. "In analyzing and designing manufacturing systems, we need to combine new organizational and managerial knowledge with that from physical and operational systems," Little observes (in this volume, p. 188). "Many of the issues involved are ill understood today and create fruitful research agendas."

Although the subsequent discussions are addressed to all readers concerned with the competitive position of the U.S. manufacturing enterprise, it should be noted that no attempt has been made to provide immediate operational solutions to today's manufacturing problems. Such actions would, of necessity, be specific to an industry, a firm, or a plant. This does not suggest, however, that every plant or firm must treat its problems as if they were unique or unprecedented. Nor does it imply that principles of operation are too general to be useful in the daily operation of the manufacturing system. The complex operational problems and the numerous interdependencies among the functions that make up the manufacturing system present a challenge to all who aspire to improve its performance. Manufacturers must constantly remind themselves that a systems perspective of the manufacturing enterprise is critical to accomplishing the desired objectives.

3

Management Practice

> Broad-ranging innovative changes in U.S. manufacturing will require the same kind of farsighted risk-taking leadership that earlier made U.S. industry the envy of the world. Without such vision all the national policy shifts and opportunity potentials imaginable will come to naught.
>
> —James Brian Quinn, NAE 18th Annual Meeting, 1982.

The complexity of the manufacturing enterprise has been noted repeatedly—the processes, the interactions among elements inside and outside the enterprise, and the impact that the environment and the customer have on the manufacturing system. The sometimes disparate and conflicting interests represented by these many elements can be both perplexing and frustrating to the management of the enterprise. Badore (in this volume, pp. 85–86) describes these circumstances well:

> The customer was not recognized as having the determining influence on product attributes or performance; tensions existed among various units and between various levels of the enterprise; no mechanism existed for setting priorities among the many desirable corporate objectives; suppliers were treated as a necessary evil to be tolerated but not trusted; and management and labor were confrontational in attitudes and objectives.

How is management to balance the constraints of capital investments against the various demands for faster response to the marketplace with new or improved products? How are they to react to demands by employees for a greater voice in operations while at the same time responding to the desires of middle management for stronger control over the system? How do they set their goals and objectives in a rapidly changing environment, and how do they know whether they are progressing rapidly enough to meet them?

There are no simple or unique answers to these or the myriad other equally critical questions that confront management daily. There are, however, some foundations that have been tested and found to be important guides to management in searching for the proper answers.

GOALS AND OBJECTIVES

Few manufacturers have the luxury of serving a market that is free from competition. Whether that competitor is local, national, or international is frequently of little significance. If a competitor is capable of producing a product that has greater appeal to the customer at a price that is attractive, the chances are good that a means will be found to enter the market and to compete with you for your share of that market. Similarly, if one manufacturer wishes to enter another manufacturer's market, it is possible to do so only with a competitive product.

What then should be a manufacturer's goal if it is to guard against this possible encroachment on the marketplace? This goal is often referred to as being a "world-class" manufacturer, a term used to convey the sense of excelling. The Japanese describe it as striving to be the best-of-the-best. As Hanson observes (in this volume, p. 164), "World-class manufacturers will be recognized by the leadership they provide in attacking and resolving complex customer problems." World-class manufacturers will be in a position to offer their customers some combination—or all—of a set of product attributes equal to or better than those of any other manufacturer in the world at a price that is attractive to those customers. Excelling in any single area, such as cost, is never sufficient to guarantee world-class status. A manufacturer must be prepared to excel simultaneously in many ways, including the following:

- Lowest cost
- Highest quality
- Greatest dependability and flexibility
- Best service
- Fastest response to customer demands

To be competitive in the world marketplace, a manufacturer must first quantify the levels of performance, defined here in broad terms, that determine "world class." A company does this by benchmarking itself against its current competitors and estimating, to the extent possible, the capability of potential new competitors (Compton, in this volume). Various metrics must be evaluated to obtain a thorough appraisal of a company's performance versus that of its competition. A variety of detailed metrics are possible in the following categories:

- Financial metrics
- Product performance metrics
- Unit operation metrics
- System operation metrics
- Aggregated measures of performance

To ensure the ability to remain competitive for the long term, it is also necessary that a manufacturer determine its ability to make use of technology and the ability of its employees to respond to a changing environment. This latter characteristic is denoted by the term *employee capability*. The task of benchmarking is time consuming and often complex. It is not something that can be done once and then set aside. A company must expend a continuous effort to know the capabilities of its competitors.

Understanding a competitor's capability is but the first step in becoming competitive. Incorporating this information into the long-term goals of the organization—establishing a vision for the enterprise—is critical in achieving improved performance. Hanson (in this volume, p. 161) believes that when people understand this vision and have the appropriate information, resources, and responsibility, they will "do the right thing." Doing the right thing is based on the appropriate frame of reference and a clear understanding of the task and its scope (see also Nonaka, 1988).

For this reason, management must view goals and objectives in both the long term and the short term (see Haas, 1987). The long-term goals focus on the customer and the markets a company is prepared to enter. Short-term goals often involve operational objectives that are best established with appropriate input from knowledgeable employees. Not only must these goals be clearly and regularly communicated to all employees, but management must put in place the means by which the performance of the organization can be continuously measured against these goals. This demands that the appropriate metrics be developed and that an assessment of progress against these metrics be continuously undertaken (Dixon et al., 1990, and Johnson and Kaplan, 1987).

It is often argued, however, that the goals and objectives that are established by the enterprise are adversely affected by outside influences, in particular by the short-term influence that is exercised by the financial and investment communities. While Fisher (in this volume, p. 138) agrees that such short-term influences exist, he notes that "for management to give any more attention than necessary to short-range buyers makes about as much sense as it would for a construction company to use blocks of ice to build a bridge across a river in the tropics." Although this book focuses largely on long-range objectives and goals, a company must not lose sight of the importance of short-term objectives.

> FOUNDATION: World-class manufacturers have established as an operating goal that they will be world class. They assess their performance by benchmarking themselves against their competition and against other world-class operational functions, even in other industries. They use this information to establish organizational goals and objectives, which they communicate to all members of the enterprise, and they continuously measure and assess the performance of the system against these objectives and regularly assess the appropriateness of the objectives to attaining world-class status.

THE CUSTOMER

A manufacturing organization serves a variety of customers. In addition to the customers who expect to purchase high-quality products and services, the owners or stockholders may also be thought of as customers in that they expect a reasonable return on the investment that they have made in the company. The employees are customers in that they expect an employer to recognize their contribution to the success of the company and to provide them with a reasonable reward for their efforts. These are the stakeholders in the organization in that each has made a personal commitment to its success. The stakeholders have special expectations and needs that must be met (Peters, 1987).

These needs cannot be met, however, unless the organization recognizes that it and its various subelements exist to provide a product or service that someone wants and is willing to pay for. As Hanson points out (in this volume, p. 160):

> Customers do not buy manufacturing, engineering, or sales; they buy solutions that fill needs. The successful manufacturer will focus the organization on customer needs, not on the functional capabilities of the organization. In this way the entire enterprise is optimized around meeting the customers' needs, using the skills of each discipline, focusing on the real task, and ultimately solving the real problems.

There are, of course, customers both inside and outside the organization. Outside are those who purchase the product or service. Inside are customers who use the products or services of other groups as they work to provide the product or service to the outside customers. Identifying the customer and ensuring that the organization is properly focused on the "true" customer is critical.

As Edmondson points out (in this volume), identifying the customer and then identifying the true wants and needs of that customer may be difficult. He argues forcefully that unless this task is carried out carefully and objectively, the enterprise may find itself providing a product or service that no one wants or needs. Edmondson suggests that one useful way to approach the task of identifying customers and their needs is to adopt the objective of helping customers meet their goals rather than providing for customers' wants. He argues that this approach will not only require a manufacturer to search to fulfill the wants and needs of the customer but will also create a frame of mind that leads to providing much more imaginative products and services for customers (further discussion about linking product characteristics and customer requirements is found in Hauser and Clausing, 1988).

Management has a special responsibility to lead and encourage the organization; in turn, all the members of the organization have the responsibility to focus on the customer. Focusing on the customer is so important that it must be treated as a foundation of good management. One cannot assume that the obvious nature of this foundation ensures that everyone in the organization understands or accepts it or that everyone shares a common focus. Manufacturers need only remind themselves of the shock that many firms have experienced when their customers chose a competitor's higher-quality products even though the price was higher.

> FOUNDATION: World-class manufacturers instill and constantly reinforce within the organization the principle that the system and everyone in it must know their customers and must seek to satisfy the needs and wants of customers and other stakeholders.

THE ORGANIZATION

The complexity of the manufacturing system arises from many directions: the interdependence of the elements of the system, the influence of external forces on it, the impact that it can have on its environment, and the lack of predictability in the consequences of actions. The complexity and the difficulty in assessing the directions that should be followed can create a sense of frustration and futility for the management.

In focusing on the systems that create, assemble, test, and service products, it is necessary to recognize that individual manufacturing operations

Japanese Management and Technology

The one great economic power to emerge in this century—Japan—has not been a technological pioneer in any area. Its ascendancy rests squarely on leadership in management. The Japanese understood the lessons of America's managerial achievement during World War II more clearly than we did ourselves—especially with respect to managing people as a resource rather than as a cost. As a result, they adapted the West's new "social technology"—management—to make it fit their own values and traditions. They adopted (and adapted) organization theory to become the most thorough practitioners of decentralization in the world. (Pre-World War II Japan had been completely centralized.) And they began to practice marketing when most American companies were still only preaching it.

Japan also understood sooner than other countries that management and technology together had changed the economic landscape. The mechanical model of organization and technology, which came into being at the end of the seventeenth century when an obscure French physicist, Denis Papin, designed a prototypical steam engine, came to an end in 1945, when the first atomic bomb exploded and the first computer went on line. Since then, the model for both technology and organizations has been a biological one—interdependent, knowledge intensive, and organized by the flow of information.

SOURCE: Drucker (1988b).

differ and depend on each other in ways that may not be completely understood. Attempts to describe the system in all of its complexity have resulted in a variety of approaches. Some descriptions concentrate on either the equilibrium state or the dynamics of the system, some treat the system as an assemblage of independent "black-box entities," and others emphasize the relationships among the various elements.

A proper coupling among the diverse units demands that each have an awareness and understanding of the objectives and capabilities of the other. This has not always been achieved. As Lardner observes (in this volume, p. 177),

> The general lack of satisfactory data and information management systems has encouraged the fractionalization of manufacturing. A manufacturing organization must react continually to changes in product requirements, product mix, product design, process design, material specifications, competitive pressures, and on and on with only brief periods of relative stabil-

> ity. Because of inadequate overall data and information management systems, functional groups have developed local systems in an attempt to maintain control over their limited areas of responsibility. Since objectives and values vary from group to group, and there is little or no understanding of how the actions of one group will affect all the other groups, individual group response to the changes in the manufacturing environment varies greatly. It is almost by accident that group actions are directed toward optimization of the whole manufacturing effort.

Hanson (in this volume, p. 158) argues that all of the internal and external organizations "must be integrated into a cohesive 'enterprise' working toward shared objectives. It is this *Integrated Enterprise* that allows both the manufacturer and the customer to be successful." Hanson identifies the following issues that must be addressed in order to achieve the integrated enterprise:

- Both management and employees must view themselves from the perspective of the tasks that must be accomplished, rather than from the perspective of the organization of which they are members.
- The enterprise must develop approaches that will lead to successful team orientations.
- The enterprise must be organized to focus on the needs of the customer rather than on functional structure.
- A clear set of values for the organization must be articulated.

Cook (in this volume) proposes that organizational structure has a profound impact on the capability of the enterprise to provide cost-effective high-quality products that meet the expectations and needs of the customer. He points out that the cultural order—the informal organization arising from personal relationships and shared values—in most functional organizations tends to support a sequential approach to product realization. Recognizing that the emerging paradigm for product realization is simultaneous or concurrent engineering representing full consideration of the design, engineering, manufacturing, procurement and service requirements for a new product as it evolves from initial concept to production, Cook examines alternative organizations that will support the new paradigm. He concludes that a more appropriate organization for achieving these objectives is a system/subsystem organization, in which the system unit has "the chief responsibility . . . to understand the customer's changing needs and translate them into a set of specifications for each individual subsystem."

Cook states that in this organization, in contrast to the functional organization, authority and responsibility are coterminous, coequal, and clearly defined. As is discussed by Badore, Hanson, and Welliver (in this volume),

Characteristics of Time-based Companies

What distinguishes a time-based organization from a traditional one? Basically, it has asked two simple questions. What deliverables do my customers want? And, what organization and work process inside my company will most directly provide these deliverables? With the answers to these questions in hand, the time-based firm then shapes its operations and policies.

How Work is Structured
Time-based companies approach work differently than do traditional ones. People in time-based—or fast-cycle—companies think of themselves as part of an integrated system, a linked chain of operations and decision-making points that continuously delivers to customers.

Traditional Companies	*Time-based Companies*
Improve function-by-function	Focus on the whole system
Work in departments, batches	Generate a continuous flow of work
Invest to reduce cost	Invest to reduce time
De-bottleneck to speed work	Change upstream practice to relieve down-stream symptom

Time-based Performance Measures
Time-based companies go back to basics when they decide how they are going to keep track of their performance. They use time-based metrics as diagnostic tools throughout the company and set basic goals of operation around them. . . . They use time to help them design how the organization should work.

Traditional Companies	*Time-based Companies*
Cost is the metric	Time is the metric
Look to financial results	Look first to physical results
Utilization-oriented measures	Throughput-oriented measures
Individualized or department	Team measures

SOURCE: Stalk and Hout (1990).

this placing of responsibility and authority with the individual—the empowerment of the individual employee—is critical to accomplishing the objective of continuous improvement.

As Mize notes (this volume, p. 197), "Increasingly, managers will have to visualize their businesses and organizations at a point in the future, interpolate their way backward into the current reality, and then aggressively manage the implementation of the transition path from here to there. But the future vision is a moving target, and the backward interpolation process must be ongoing and dynamic." Mize points out that the enterprise and the organization that is established to support it are dynamic and must be constantly subject to review and, when appropriate, to change. It is only by this means that they will remain capable of meeting the current challenge and prepared for the future challenge, in short, that they will have the means and the desire to renew themselves continuously in view of the changing events in the world marketplace.

> FOUNDATION: A world-class manufacturer integrates all elements of the manufacturing system to satisfy the needs and wants of its customers in a timely and effective manner. It eliminates organizational barriers to permit improved communication and to provide high-quality products and services.

THE EMPLOYEE

Accomplishing the objective of creating a world-class manufacturing organization must begin with recognition that the most important asset of an enterprise is its employees. When properly challenged, informed, integrated, and empowered, the employees can be a powerful force in achieving the goals and objectives of the organization. People are key to achieving a world-class competitive status; they are absolutely essential for success, although they alone cannot ensure success (see also, Prahalad and Hamel, 1990, and Senge, 1990).

Creation of the environment in which employees can participate in the activities of the organization demands a change in the thinking of many people. It is not just the supervisors, managers, vice presidents, the president, and the chairman that must be willing to participate, but also the employees on the plant floor. Employee involvement, as Badore notes (in this volume), means including employees in the operation of the firm. This involvement has two objectives:

- Creating and sharing of the vision of goals and objectives—for the overall enterprise as well as for each organizational unit—by all employees.
- Seeking and sharing the knowledge possessed by the individual employees to identify and solve problems.

The rationale for employee involvement is predicated on the assumption that individual employees have the best opportunity to understand and appreciate the problems that are unique to their positions. They know their jobs and they know what limits their performance.

The employee involvement strategy is directed at letting employees decide the best way to do their jobs. Marsing (in this volume, p. 190) observes that "For senior managers to sit in their offices and assume that they have all the knowledge and experience needed to make critical risk decisions is a prescription for failure. To make the best decisions with the most knowledge, managers have to use the combined intelligence of the entire work force." Hanson (in this volume) emphasizes this as one of the key principles in achieving an integrated enterprise. Welliver (in this volume) refers to the process of going to where the information resides—in the employees—as "going to the gemba."

Badore (in this volume, p. 87) emphasizes the importance of employee involvement.

> [It] is, in a sense, the means by which a large organization attempts to achieve many of the benefits that are generic to the small organization. Although certain organizational structures and systems are required in larger organizations, the effort to accomplish meaningful employee involvement . . . is directed at preventing the organizational structure and systems from providing barriers to finding the best solutions to problems. Furthermore, the involvement process provides a means of humanizing the organization and maintaining participation by individuals at all levels—a process that is intended to lift the organization to new heights of performance through the best use of the skills and interests of the individual. It is the means by which continuous improvement can be made an operating goal for all levels of an organization.

Wilson (in this volume, pp. 239–240) uses the experience of the jazz musician as an example of interactions that are important for group success.

> In small group improvisation (fewer that eight players), the players must share a commitment to excellence demonstrated through the creativity and imagination of their improvisations. The group conditions must free the players to establish the group cohesion and interdependence of their own contributions. To achieve group excellence, each of the players must be highly skilled. . . . Communication among the players is essential during performance. . . . They intimately share instant information about their performance, have the power to determine and modify its direction, share full knowledge of the performance technology, and immediately share the rewards of the audience response. . . .
>
> Thus, the concept of group creativity in a jazz performance may provoke some new ideas for organizing production work. For example, the moments of creative opportunity for a jazz musician are a small but highly motivating fraction of his total professional life. Hours of uncompensated practice are required to achieve those creative moments. Would most employees be similarly motivated by the opportunity for occasional breaks from routine work to engage in a creative job experience?"

Employee involvement does not, by itself, provide the mechanism by which employees can use their knowledge and experience to benefit the enterprise. Badore notes that "If proper advantage is to be taken of the knowledge that the employee possesses, it is necessary to empower the employee to implement the solutions that they know to be available. By so doing, the enterprise is making the employee an integral part of the solution process." Hanson argues that creating the integrated enterprise requires "empowerment of the individual" and that this leads to "distributed decision making [because] information freely shared with empowered people who are motivated to make decisions will naturally distribute the decision-making process throughout the entire organization."

Fisher brings the perspective of the financial community to this matter of continuous improvement. He says (in this volume, p. 142), "For my own investments and those I handle for others, I am interested only in companies that recognize that competition is steadily improving, so that it is incumbent on these companies continuously to improve their own efficiency and never to be satisfied even with the quite magnificent strides that some of them have made in recent years."

> FOUNDATION: Employee involvement and empowerment are recognized by world-class manufacturers as critical to achieving continuous improvement in all elements of the manufacturing system. Management's opportunity to ensure the continuity of organizational development and renewal comes primarily through the involvement of the employee.

THE SUPPLIER OR VENDOR

Figure 1 explicitly recognizes the importance of the role of suppliers and vendors to the integrated manufacturing system, a role that has long been recognized by Japanese manufacturers as critical to their success. The close relationship between supplier and purchaser in Japan has created a situation that "represents a form of vertical integration without the actual legal or direct financial commitment that would be required of 'true' integration" (National Academy of Engineering, 1991, pp. 101–102). A similar type of long-term relationship between supplier and purchaser is being created in this country, in recognition of the fact that the system of manufacturing encompasses these elements as well. As Hanson notes (in this volume), "A supplier unfamiliar with marketing plans and product strategies cannot fully provide the resources and intelligence to help reduce time to market."

It is essential that the barriers that have existed between supplier and purchaser be attacked as actively as are the barriers between the elements in the manufacturing organization, for example, product design and process design. The sharing of goals, the exchange of information, the interchange of people, and the making of long-term commitments are some of the ways in which these barriers are being overcome (further discussion of supplier relationships can be found in Womack et al., 1990).

Essentially everything that has been said in this report—both above and in much of what follows—is as applicable to the supplier as it is to the manufacturer. The successful manufacturing system will create an environment that encourages, recognizes, and rewards the integrated involvement of all elements of the system.

> FOUNDATION: A world-class manufacturer encourages and motivates its suppliers and vendors to become coequals with the other elements of the manufacturing system. This demands a commitment and an expenditure of effort by all elements of the system to ensure their proper integration.

THE MANAGEMENT TASK

Effective management is critical if an enterprise is to compete in the world marketplace. The committee has identified five foundations that relate to management practice—establishing the goal of being world class, attending to the needs and wants of the customer, creating an effective organization, creating an environment that encourages and rewards employee involvement and fosters employee empowerment, and integrating the suppliers and vendors into the system.

Attention to any one of these will be useful to an organization, but achieving world-class status will require that all five be simultaneously pursued. The challenge to management is to understand the importance of the task, to commit to accomplishing it, and to devote the enormous effort that is required to complete it.

Welliver (in this volume, p. 237) describes the task as follows:

> Our biggest challenge is to instill this philosophy and approach throughout the company. Managers are accustomed to having more control over the processes, but they do not realize they will have much more control over the quality of their products or services if they let go of some of the decision making. . . . We are also learning that managers must look for problems with vigilance, gaining an understanding of the real issues and problems by poring over facts and data. . . . But once the hidden problems are revealed, managers start to realize that the appearance of a smoothly operating organization can be deceptive.

Malcolm Baldrige National Quality Award

The Malcolm Baldrige National Quality Award is an annual Award to recognize U.S. companies that excel in quality achievement and quality management. Fundamental to the success of the Award in improving quality in the United States is building an active partnership between the private sector and government.

Companies participating in the Award process submit applications that provide sufficient information and data on their quality processes and quality improvement to demonstrate that the applicant's approaches could be replicated or adapted by other companies.

Key Concepts in the Award Examination Criteria

- Quality is defined by the customer.
- The senior leadership of businesses needs to create clear quality values and build the values into the way the company operates.
- Quality excellence derives from well-designed and well-executed systems and processes.
- Continuous improvement must be part of the management of all systems and processes.
- Companies need to develop goals, as well as strategic and operational plans to achieve quality leadership.
- Shortening the response time of all operations and processes of the company needs to be part of the quality improvement effort.
- Operations and decisions of the company need to be based upon facts and data.
- All employees must be suitably trained and developed and involved in quality activities.
- Design quality and defect and error prevention should be major elements of the quality system.
- Companies need to communicate quality requirements to suppliers and work to elevate supplier quality performance.

Although reliable evaluation relative to the Examination criteria requires considerable experience with quality systems, the Examination may also be used for self-assessment and other purposes. Thousands of organizations—businesses, government, health care, and education—whether or not they are currently eligible or plan to apply for Awards, are using the Examination for training, self-assessment, quality system development, quality improvement, and strategic planning.

SOURCE: National Institute of Standards and Technology (1991).

Hanson (in this volume, p. 164) elaborates on that theme:

> The successful manager of the 1990s will have the skills to define complex dilemmas and resolve them, not ignore them. This management skill, which can be defined as 'dilemma management,' is a critical component of the Integrated Enterprise. The characteristics of the dilemma manager include the ability to tolerate ambiguity, to manage and, indeed, thrive on the tension that is caused by apparently conflicting demands. The apparent conflict will be valued as a stimulator for change.

Marsing (in this volume, p. 191) suggests that the role of senior management is to understand and remove obstacles that impede the progress of operational units. "Anything less will not build the foundation needed in the organization to deal with change and risk taking."

The task facing management is indeed daunting. It is tempting to ask whether there is a single "right way" to go about addressing this task. The answer is probably not. The approaches depend on many circumstances, including such things as the personalities of the people involved, the size of the organization, the level of competition in the particular industry, and the rate of change that the industry is undergoing. Wilson (in this volume, p. 241), in his use of the jazz group as a metaphor for the manufacturing enterprise, notes that the nature of the leadership varies with size of the group.

> As the number of musicians in a jazz ensemble increases, collective improvisation becomes increasing hard to execute. . . . [L]arger jazz ensembles use written arrangements. . . . The 'big band,' comprising 12 or more players, removes much of the self-determination from the individual player, limiting his creative contribution. . . . Those 'big bands' that survived through several eras were distinguished by a single leader . . . who established an identifiable sound for the group.

Effective leadership can be achieved in many ways. Again Wilson's observations concerning leadership of jazz groups seems equally applicable to the manufacturing enterprise.

> In contrast to small jazz groups, the leader of a big band has a strong individual role in establishing the style and the expectations about the quality of performance. He relies on written communications (arrangements) to provide the structure of the performance relationships and to indicate where individuals can contribute their own creativity through solos. Nevertheless, the anecdotes suggest how widely the leadership styles of the band leaders may differ. Although the style and discipline of the band may therefore be a reflection of the leader's personality, there seems to be no obvious correlation between leadership style and commercial success.

Leadership, Incentives, Rewards

Companies produce fine products largely because the people at the top care about the product per se, elevate product or innovative people to strategic levels, and commit resources behind them. Company managements that look at technology or manufacturing activities simply as money mills to be compared against the financial advantages or disadvantages of hoarding silver or owning banks are unlikely to create the internal pressures or atmosphere that keep their organizations strong, processes current, quality high, and technologies at the forefront. Financial measures rarely reflect these crucial aspects of performance until years after the most critical actions have been taken or ignored. Sony has been an innovative leader because Messrs. Ibuka and Morita are talented and have long cherished innovation and quality products per se. Pilkington's float glass innovations occurred because Alastair Pilkington wanted to invent, and its top management had long time horizons, understood the need for innovation, and empathized with the chaos and risks involved. Japan has emerged largely because its leaders had vision, patience, and a high regard both for technological advance and for building the worth of their human resources.

Until boards appoint and reward top managers for being innovation oriented and interested in the company's future product and cost positions, U.S. manufacturing companies and industries will suffer. Fortunately, when plans are well conceived and communicated, the stock market does reward progressive companies with high P/E ratios, the basic method of allocating less expensive capital in the United States. To be effective, this longer-term focus must also be reflected in the full control and reward systems of the company. Properly developed, multiple goal "management-by-objectives" (MBO) systems, combined with carefully designed strategic portfolio plans and controls, provide available mechanisms for orienting lower-level decisions toward the future. Unfortunately, too few companies use these mechanisms to their full capability, relying mostly on short-term accounting and return on investment (ROI) controls instead. Smaller companies often have longer-term horizons because their owner-managers look to future stock market yields rather than to more current rewards. A greater use of measures and rewards that generously compensate large company executives for their units' total performance five years later might engender very useful effects.

SOURCE: Quinn (1982).

Imaginative, creative leadership at every level of an organization is critical if it is to be capable of building on these foundations. Management creates the culture within which the organization functions. Management must exhibit the concern for the health and well-being of the organization's human resources. Management must insist that the organization look beyond its borders to interact with its customers, its suppliers, and the educational systems that are training its present and future employees. It is a challenge to the organization to find the proper management for the circumstances in which it finds itself (see also Drucker, 1990, and Mintzberg, 1978).

> FOUNDATION: Management is responsible for a manufacturing organization's becoming world class and for creating a corporate culture committed to the customer, to employee involvement and empowerment, and to the objective of achieving continuous improvement. A personal commitment and involvement by management is critical to success.

4

Measuring, Describing, and Predicting System Performance

> When you can measure what you are speaking about, and express it in numbers, you know something about it, . . . (otherwise) your knowledge is of a meager and unsatisfactory kind; it may be the beginning of knowledge, but you have scarcely in thought advanced to the stage of science.
>
> —Lord Kelvin (1824–1907)

Operating a business enterprise in such a way that objectives are achieved in the most efficient and timely fashion requires the optimization of system performance. One is seeking to maximize or minimize multiobjective functions within a specified set of constraints; for example, maximizing profits, or minimizing capital expenditures, defects, or material use per unit of product, within such constraints as fixed total resources, equipment configuration, or product mix.

Determining an optimal strategy for a complex manufacturing system, whether it is for control, investment, or processing, is seldom straightforward. Since the system often consists of elements that respond in a nonlinear way to inputs provided by other elements of the system, one must understand the detailed interactions that exist if one is to optimize the total. Considering this complexity, the operational approach is frequently to decompose the system into supposedly independent subsystems, such as areas, shops, cells, or units, and then to optimize the performance of each subsystem and impose interactions among the subsystems in such a way that an overall optimum is obtained. The intent is to arrive, through repetitions of this process, at a solution that properly optimizes the original system.

Despite the difficulties that can arise in this procedure, it is important to recognize that there may be few alternatives to its use. The alternatives are limited for the following reasons:

1. Many manufacturing systems are too large to treat as a single entity.
2. The interactions among the various subsystems of a manufacturing system are frequently nonlinear.
3. The behavior of some subsystems cannot be described mathematically.

In formulating "foundations" of manufacturing systems, it is important to understand the value and the limitations of this piecewise approach to determining optimal operating strategies. While the insight that these analyses provide can be of great value in understanding many aspects of the manufacturing system, it is important to recognize the difficulty in determining the value of any given solution. This limitation, however, does not diminish in any way the importance of using models and mathematical constructs to provide insight into the performance of the manufacturing system.

METRICS: QUANTIFYING THE PERFORMANCE OF THE MANUFACTURING ENTERPRISE

Performance evaluation is a process applied throughout the manufacturing enterprise to measure the effectiveness in achieving its goals. Because of the variety, complexity, and interdependencies found in the collection of unit processes and subsystems that define the manufacturing system, appropriate means are needed to describe and quantify rigorously the performance of each activity. Metrics are mechanisms used in describing and appraising those systems, subsystems, and elements (see Dixon et al., 1990, and Johnson and Kaplan, 1987).

Manufacturers have three basic sources for metrics; many are part of general business knowledge and are readily available in the management literature, especially the metrics used to describe and evaluate the financial performance of the firm. A second group might be characterized as industry specific metrics. They are commonly recognized as appropriate measures of some aspects of the manufacturing system, usually within a single discipline; for instance, metal forming. And finally there are metrics that are developed by individual companies reflecting their special circumstances. Metrics developed for these specific contexts can provide the basis for a period of unique competitive advantage, although these metrics will be broadly adopted within a particular industry and later within the general pool of recognized manufacturing wisdom. For instance, lean production (Womack et al., 1990) and just-in-time inventory control (JIT) both depend upon the development and use of appropriate metrics. In the case of lean production, these metrics are used to measure the amount of resources consumed in design and production of products, driving toward a minimum. JIT metrics focus on the time between entry of materials into inventory and their incorporation in products on the production line. Although both of these metrics were refined within the automotive industry, they are becoming widely adopted for manufacturing in general.

Taxonomy for Metrics

An orderly means of classification should provide some initial help in the selection and use of suitable metrics; knowing where to find them, however, offers only a minimal basis for understanding which to use and when their use is appropriate. Cook (in this volume) suggests a reasonable taxonomy can be developed with a simple division of all metrics into direct and proxy metrics. Direct metrics have no intervening transformation between the measures of a variable and its associated value; the correlation is immediate. For instance, if the number of product defects is indicative of the quality of a process, it would be expected that if the number of product defects declines the quality of the process has increased.

Proxy metrics, however, involve transformation of the values of several variables to arrive at the value of the enterprise performance metric. They are often complex aggregations of many, possibly diverse, characteristics that may not directly influence characteristics of interest in the organization. For example, patent activity is a proxy measurement sometimes used as an indicator of the innovativeness of a manufacturer (see Howard and Guile, 1992). Profitability is a proxy metric that a manufacturer may use to gauge performance. However, it is not usually possible to influence profitability by adjusting one or two quantities or characteristics. Therefore, proxy metrics may be better thought of as indicators of change rather than cause-and-effect relationships that are directly manipulable.

The rules and policies that guide decisions at many different levels of the manufacturing company must incorporate the metrics appropriate to each. At the corporation level, proxy metrics are often used for measuring the profitability and customer responsiveness of the entire enterprise. For the subsystems there are additional metrics, such as yield from a series of processes, that reflect the performance of an integrated subsystem of production equipment and its accompanying labor force; once again these are likely to be aggregations of many variables that "indicate" the performance of the subsystems. However, at the component and unit operation level of the manufacturing firm, we find a set of direct metrics, reflecting the performance of the machine, process, or worker; for instance, the number of defects identified, the productivity of the equipment, or the number of days of employee absence. Marsing (in this volume) notes the importance of formally developing and using statistical metrics for each step of the production process, and describing the range of upper and lower bounds within which the machines and equipment are expected to perform. Then, when the performance exceeds previously defined limits, corrective action can be taken at the operator level by modifying machine settings rather than at the aggregated process level, which is less well understood.

Matching Metrics with Goals and Concerns

In the choice of metrics, manufacturers have two perspectives that should be considered; one external measure and one internal.

> First, . . . measure the performance of the organization against that of its competitors and, second, . . . assess the trends in one's performance in order to take appropriate actions to ensure continuous improvement. In the first case, an absolute measure of performance is needed. This is sometimes referred to as "benchmarking," or measuring oneself against the world leader—the "best-of-the-best" in a product or process arena. In the second, progress over time is the prime concern, that is, how well the organization is achieving continuous improvement in performance. A proper combination of these two measures is critical. Without them an organization cannot properly evaluate its absolute competitive status, nor can it be assured of its ability to remain competitive over a long period of time (Compton et al., in this volume, pp. 107–108).

A particular concern to manufacturers of complex or innovative products incorporating rapidly changing technologies is the difficulty in identifying the correct metrics for important characteristics of the product and associated processes. In the early phases of the product life cycle, gross measures of performance are likely to be appropriate or sufficient, many times because a set of clear cause-effect relationships has not been established for the new product. However, as the product and associated process mature and are better understood (more scientific foundations, better models), those metrics do not adequately reflect the advances made in the product's performance or in the production activities required for its manufacture. This indicates the need either to consider the continuing validity of those metrics being used or to expand the portfolio of metrics.

When metrics are not readily available to measure the important characteristics, then modification of available metrics or development of new bases for measurement is called for. Hewlett-Packard found it necessary to develop its own "internal" metrics for measuring the effectiveness of their

Appropriate Metrics for Rapidly Changing, High-Technology Products

When General Electric Medical Systems first began manufacturing magnetic resonance imaging (MRI) systems, the number of defects identified in each unit prior to shipment was the primary metric for internal product quality. However, as design and production engineers developed better understanding of the complex interactions among the technologically sophisticated components of the MRI products, they also increased the number of opportunities to detect defects or out-of-specification occurrences in their product. This created a paradox: as the quality of the products delivered to the customers of GE increased, the internal metric used to evaluate the production staff—the number of defects identified and corrected before shipment—indicated falling performance. Clearly a better means of reflecting the increasing quality and competence of the production system was needed.

The length of time that the order spent in the manufacturing facility—its cycle time—and the variance of the cycle time were identified as a more representative metric for the quality of the product and production processes. Cycle time focuses attention on a characteristic that remains consistent throughout successive product generations, and it can incorporate changes in process and product technologies. Therefore, it does not penalize design and production engineers for furthering their understanding of the fundamentals underlying the product and developing mechanisms for managing the idiosyncracies of its manufacture. The variance of the time that the orders spend in the facility is an important indicator of the level of control the production facility exercises over the associated manufacturing processes.

The effectiveness of this metric is illustrated by the significant decreases in product cycle time over successive generations of MRI products. When GE engineers began measuring cycle time, the unit of measure was weeks; current measures are indicated in days, with their goal cycle time measured in number of work-shifts.

SOURCE: Personal Communication, 1990. Frank Waltz, Manager of Magnetic Resonance Manufacturing, General Electric Co., Waukesha, Wisconsin.

cross-functional product development teams (House and Price, 1991). While the "return map" developed by HP contains many metrics that are similar to those discussed in this volume, the metrics have been modified, tweaked, and combined in a manner that reflects the unique measurement needs identified by HP for its collaborative product development efforts. The metrics create a unique internal language, or grammar for communication among the many different functional groups in the company (Lardner, in this volume). As described by House and Price (1991), the metrics are used to measure, communicate, and understand the "contributions of all team members to product success in terms of time and money . . . (and the return map) includes the critical elements of product development and the return or profits from that investment."

Financial Measurements

Of course the survival and success of all manufacturers—even those enjoying dominant market positions and offering the highest quality products—depend on the ability of the company to make a profit for its owners. Therefore, there is a real need for the proper financial metrics to measure, compare, and furnish information for decisions about the management of the enterprise.

Care in selection of appropriate metrics is extremely important because they focus attention on a particular set of variables and thus affect the kind and direction of control taken. Cook (in this volume) points out that U.S. manufacturers are beginning to realize that proxy financial metrics, such as profitability, market share, and return on investment have not been very good for measuring the effectiveness of manufacturing in a global marketplace. He suggests that quality, lead time, flexibility, and innovation may be better reflections of their competitiveness. These goals represent fundamental metrics that can be directly influenced by the enterprise and, in turn, improve future financial performance and resilience to change.

For some time, engineering managers have expressed concern with the difficulty of available methods for justifying capital investments in such areas as flexible manufacturing equipment, increasing product quality through better production process controls, and promoting greater work force involvement through employee education programs. The management accounting community has finally recognized these problems and begun to question the metrics incorporated in the manufacturing operating policies, control parameters, and performance-evaluation criteria that are used to evaluate the return/viability of new projects (Johnson and Kaplan, 1987). Policies based on information derived from aggregate financial reporting data (proxy metrics) offer almost no basis for operational decisions or evaluation of investments in new technologies (Eccles, 1991). Chew and coauthors (1990) show how a company with 40 plants, all producing basically the same products, missed opportunities for increasing profitable performance because when comparing the plants, division management focused on the wrong financial metrics. They considered the most effective plants to be those that were most profitable and ignored the special circumstances that caused locations with outstanding productivity to exhibit only good profitability. The transfer of the ideas, methods, and technologies from the high-productivity plants would further increase the profitability of the higher ranking profit centers.

It must also be kept in mind that the use of financial measures alone should not, as is so frequently the case, be the only source for metrics. Turnbull and his coauthors (in this volume) suggest that nonfinancial metrics be used to assess the plant, business unit, or enterprise performance and that they in turn will contribute to predicting the expected performance of the business financially. They point out that there is likely to be a wealth of data available to the planner but it requires a systematic search, often from a number of sources that have not commonly been included, such as operating staff and even outside sources. It is important to develop the correct mix of financial and nonfinancial measures to guide the manufacturing organization. Each set of indicators provides different perspectives on the manufacturing system, and one must recognize that changes in one set of metrics may not be reflected by accompanying changes in another set. For example, it may not be easy to quantify the financial returns expected from investment in computerized flexible machining systems or training the work force in quality function deploy-

Nonfinancial Indicators and Long-Term Profitability

In an important sense, a call for more extensive use of nonfinancial indicators is a call for a return to the operations-based measures that were the origin of management accounting systems. The initial goal of management accounting systems in the nineteenth-century textile firms and railroads was to provide information on the operating efficiency of these organizations. Measures such as conversion cost per yard or pound and the cost per gross-ton-mile provided easy-to-understand targets for operations managers and valuable product cost information for business managers. These measures were designed to help management, not to prepare financial statements. The need to expand summary measures beyond those used to measure the efficiency of conversion reflects the greater complexity of product and process technology in contemporary organizations. But the principle remains the same: to devise short-term performance measures that are consistent with the firm's strategy and its product and process technologies. We need to recognize the inadequacy of any single financial measure, whether earnings per share, net income growth, or ROI, to summarize the economic performance of the enterprise during short periods.

SOURCE: Johnson and Kaplan (1987).

ment. But nonfinancial metrics that highlight the effect of switching rapidly between a wide variety of production setups will readily indicate the opportunity for a significant improvement in company performance.

The Application of Metrics as Operational Guidelines

Assuming that the proper metrics have been selected, one must next determine how those metrics can be applied. This implies developing rules or policies to guide the collection of data and judging the meaning of the values found. Although the desired direction of the effect can generally be defined, qualitative measures are in general not sufficient; a quantitative means for establishing the norms for policy parameters is best. For example, Compton and coauthors (in this volume) show that learning curve models can be used to establish expectations for quality metrics and to correlate values of those metrics with specific actions taken for their improvement. In their view

> Learning curves are not to be viewed as merely descriptive. They can be, and frequently have been, used as an aid in making predictions, in that early experience in the production of a product can be used to predict future manufacturing costs. [Assuming confidence in the parameters chosen] one can readily predict the costs to produce a unit after some future cumulative production volume has been achieved.

Although these metrics are important for considering the activities within the organization over time, it is at least as important to maintain a vigilant awareness of competitors' capabilities and to adopt or define effective metrics for comparisons. Competitive strengths of the manufacturing firm have become more dependent on the quality of its work force and their ability to incorporate appropriate new technologies in products, processes, and services of the company (see also Prahalad and Hamel, 1990). Compton (in this volume) emphasizes the importance of developing the proper metrics for identifying, measuring, and evaluating these characteristics to ensure the future viability of the company. His metrics for gauging the capabilities of the organization to evaluate technological developments include the level of support for internal research and development; the portion of the R&D budget devoted to long-term projects, exploratory activities, new concepts, and technological innovations; the level of encouragement and support personnel receive to participate in worldwide technical meetings and activities; and the level of investment in technical libraries and information resources.

When assessing the capabilities of the technical work force relative to one's competition, credible indicators include the distribution of professional and advanced degrees as well as involvement in continuing education. Professional awards and participation in outside organizations (for example, as officers, speakers, and lecturers) are indicative of the quality of the employees most responsible for maintaining the technological competencies of the firm (Compton, in this volume).

Other Metrics

Although it is necessary to measure performance throughout the manufacturing enterprise, it is also necessary to apply metrics beyond the internal measures discussed above. Edmondson (in this volume) discusses the importance of using metrics that reflect whether the product definition, captured by designers and marketing staff, matches that articulated by customers. Each time there is additional information to share with the customer—presenting the design specifications, demonstrating a prototype—the metric remains the same: "Is this what you had in mind?" Cautioning against shortcuts, Edmondson points out that making use of this metric imposes considerable work on the manufacturer and also, to a lesser extent, on the customer.

> Some firms might be tempted to establish metrics that they can generate and test from within the firm itself. . . . Metrics of this sort can give some indication of how well you are meeting your product definition but certainly seem to be a poor substitute for a real, live customer reaction. . . . In the final analysis the customer's reaction to your product has a nearly 100 percent correlation with whether or not your product will sell (Edmondson, in this volume, p. 135).

> FOUNDATION: World-class manufacturers recognize the importance of metrics in helping to define the goals and performance expectations for the organization. They adopt or develop appropriate metrics to interpret and describe quantitatively the criteria used to measure the effectiveness of the manufacturing system and its many interrelated components.

MODELS AND LAWS

Laws, in the context of scientific and engineering discovery, provide intellectual foundations and explanations by describing the relationships among the variables and parameters of the phenomena under investigation. These science-based laws also make it possible to predict the consequences of changes in variables based on an understanding of the relationships among them. However, as the number of variables increases and their relationships become more complicated and less well understood, we are less certain of the effect caused by changes in one or more of the variables. As the relationship of the variables becomes more complex and the phenomena described become less specific, we begin to consider their explanations to be less reliable and more subject to outside forces or dependent on the context or environment. Eventually, the "laws" are considered as models of the situation of interest.

Many of the phenomena of nineteenth-century physics that were identified as laws of nature were, by the mid-twentieth century, spoken of as models of phenomena. Models and modeling continue to be the popular terminology, particularly as Little observes (in this volume) "in the study of complex systems, social science phenomena, and the management of operations." Little suggests that the word *model* conotes the "tentativeness and incompleteness" often appropriate to our descriptions of complex systems "in which there are fewer simple formulas, fewer universal constants, and narrower ranges of application than were achieved in many of the classical 'laws of nature'."

The goal then for metrics and models is the identification of manufacturing science-based explanations and foundations—"laws of manufacturing systems"—that could be used to describe and understand current manufacturing systems, predict the consequences of actions, and confidently initiate the actions necessary to achieve organizational goals. Little notes that we are more likely to find a taxonomy or hierarchy of "manufacturing models" that provide various degrees of generic applicability. The bases available for constructing descriptions of phenomena are limited to mathematical expressions that have no necessary relationship to the real world, physical laws, which require observation of the world and induction about the relationships among observable variables, and empirical descriptions of the world in which there are fewer simple formulas and only approximate representations for phenomena. For example, the use of elementary queueing theory by Krupka (in this volume) to represent the flow of parts and materials through production equipment and machines in the factory is an instance of mathematics without physical foundation applied to manufacturing systems (for further discussion of models and mathematical formulations applied to manufacturing systems, see Striving for Manufacturing Excellence, 1990).

Empirical Models

Models based on observations, such as relating the physical distance between pairs of researchers and the number of "messages" they exchange per week, can be applied to a broad range of engineering and managerial practices—they are generic models. When commenting on the relationship between distance and communication frequency, Little noted that while there does not appear to be a "strictly prescribed functional form or universal constant, . . . there is definitely a general shape and an experimentally determined range of parameter values." Moreover, although "the regularity of the curves can be distorted by a variety of special circumstances . . . the basic phenomenon is strong and its understanding is vital for designing buildings and organizing work teams effectively."

Compton and coauthors (in this volume) propose learning curves for quality as another example of generic model applicability. The learning curves, based on empirical observations, have shown a general relevancy across a diverse range of manufactured products. These models for projecting quality improvement share a similar form with the models developed in the late 1930s to explain the significant decreases in product unit costs as a consequence of accumulated production volume. Several possible reasons can be proposed for their comparable configurations:

> Although the specific actions taken to improve quality differ from those taken to reduce unit costs, a striking similarity exists between the two. . . . Both result from conscious actions taken by management and employees to accomplish a common strategic objective for the enterprise. Both combine human commitment and training with technical improvements. Both require extensive knowledge of the processes being employed and the products being produced. Therefore, quality and costs might be expected to share a common representation (Compton et al., in this volume, pp. 110–111).

Although the applicability of models in the manufacturing enterprise is most commonly thought of in the context of production operations and processes closely associated with the physical manufacture of products, Krupka suggests that models and modeling should be considered in a much broader context. An example is the use of models to discover problems in the subsystems involved in new product introduction and other nonmanufacturing steps, before the start of physical production activities. Krupka notes that nonmanufacturing operations "are often more complex than those encountered on the factory floor." Moreover

> Analysis of such [models] often reveals the presence of steps that add no value or that consist of re-creating—at the risk of introducing errors—information created elsewhere. Eliminating these steps will shorten the system's interval, reduce costs, and often improve quality by reducing opportunities for the introduction of errors (Krupka, in this volume, p. 168).

Another reflection of the intricacies of the systems that operate within the manufacturing enterprise but may be hidden until observations are described with empirically derived charts is evidenced in "decision-expenditure" curves (Bowen, in this volume). The formal and informal linkages and delays that occur as new products and processes are commercialized result in long feedback loops. These empirical observations suggest the magnitude and value of up-front knowledge when time is of the essence. The startling issue is that often 80–85 percent of the project expenditures are determined during the first 15 percent of the project time. Therefore, a high priority is placed on starting research efforts in the earliest phases of the project, because, in Bowen's view, the people involved in these projects "perceive the actual time of the decisions that triggered the expenditures as occurring very early in the process."

> Complex processes involving numerous variables and elements of subsystems (such as information, technology, human, financial, and marketing) result in longer than anticipated execution consequences and, thus, strongly influence feedback loops in the manufacturing system (Bowen, in this volume, p. 94).

Bowen proposes that the important aspect of these models is the circumstances they suggest rather than the specific values indicated. Furthermore, when looking at the cases involving the "best-of-the-best," the 15/85 rule does not seem to apply. Bowen points out that in those cases, "the decisions are much more closely linked to the doing and the expenditures," and that "the feedback and corrections are different in number, timing, and quality." He suggests the following explanations for the 15/85 trends:

- The ineffective working of teams pulled from functional groups,
- The lack of standards or a single data set,
- Procedures and mechanisms for problem solving and structuring of the solutions, and
- Infrastructural issues such as lengthy procedures and justification for obtaining resources—people or capital.

Examination and attention to these relationships should be helpful in establishing proper goals and expectations and understanding how they can be facilitated.

Modeling and Understanding

The complexity, nonlinearity, and stochastic nature of models are reflected in the number of variables they include, the number of relationships and interdependencies described, and the amount of information that the model generates. The degree of complexity is, to a significant extent then, a matter of how detailed the model representation is. And the complexity will directly influence the comprehension of the model and the acceptability of the results obtained, as well as increase the difficulty of changing and enhancing the model.

Solberg (in this volume) proposes simplicity as a key to the acceptance and use of models. He emphasizes the importance of the relationship between the credibility of models and user understanding of those models and what they depict.

> It is neither necessary nor desirable to build complicated models to deal with complicated situations. Indeed, we should be trying to find a point of view that makes complicated situations seem simple. . . . We must be aware that simple does not mean trivial or obvious. We cannot define relations arbitrarily, make capricious assumptions, or generalize recklessly. . . . Finding the adequate level of detail, the appropriate assumptions, and the elegant formulation is a matter of hard work and inspired wisdom (and perhaps a large dose of luck) (Solberg, in this volume, p. 218).

And when there are several alternative representations of a particular system, Compton too advises that although determining the most appropriate model depends upon many factors, such as the data sampling protocol, it is probably best to use "the simplest formulation possible." But often the simple representation can be discovered only after constructing and examining more complex models to gain additional insight into the problem; when constructing a first model it may be difficult to determine which variables and relationships define or constrain the performance of the problem. Therefore, the first model will include many more factors than will ultimately be needed (Pritsker, 1986b).

Of course, the opportunity to compare the results of several representations can offer further assurances that the results of the models are valid. In fact, Little encourages the development of a "validity-check" model:

> If the results of running a complex model suggest a particular course of action, it is imperative to know why the model produced those results, that is, what were the key assumptions and parameter values that made things come out as they did. . . . We should have a simple model that uses a few key variables to boil down the essence of why the recommendations make sense (Little, in this volume, p. 186).

Determining Limits for Improvement

Manufacturers must be able to establish realistic goals and to plan for their accomplishment in the face of uncertainty.

> The future success of a business will be influenced both by processes over which the business has little control and by those it can affect directly. For a process in the former category, we are interested in forecasting its expected performance over time. For a process in the latter category, we are interested in forecasting its potential performance, based on our understanding of "what could be" and our capacity to act (Turnbull et al., in this volume, p. 226).

A Taxonomy of Manufacturing Process Technologies

In selecting from among alternative process technologies, a combination of materials science, mechanical, and economic analyses are used. The desired physical characteristics of the product also impose severe constraints on this selection. The following table is a sample of the variety of material conversion technologies available for changing the physical properties or appearance of materials or combining them.

Processes for Changing Physical Properties

Chemical reactions	Hot working	Heat treatment
Refining/extraction	Cold working	Shot peening

Processes for Changing the Shape of Materials

Casting	Piercing	Torch cutting
Forging	Swaging	Explosive forming
Extruding	Bending	Electrohydraulic forming
Rolling	Shearing	Magnetic forming
Drawing	Spinning	Electroforming
Squeezing	Stretch forming	Powder metal forming
Crushing	Roll forming	Plastics molding

Processes for Machining Parts to a Fixed Dimension

Traditional chip removal processes

Turning	Sawing	Boring
Planing	Broaching	Reaming
Shaping	Milling	Hobbing
Drilling	Grinding	Routing

Nontraditional machining processes

Ultrasonic	Chem-milling	Optical laser
Electrical discharge	Abrasive jet cutting	Electrochemical
Electro-arc	Electron beam	Plasma arc

Processes for Obtaining a Surface Finish

Polishing	Superfinishing	Honing
Abrasive belt grinding	Metal spraying	Lapping
Barrel tumbling	Inorganic coatings	Anodizing
Electroplating	Parkerizing	Sheradizing

Processes for Joining Parts or Materials

Welding	Pressing	Sintering
Soldering	Riveting	Plugging
Brazing	Screw fastening	Adhesive joining

SOURCE: Amstead et al. (1977).

If it is not possible to understand and describe adequately the potentials and limitations of the firm's capabilities, goals for improvement will have little rational basis for those who are charged with accomplishing them.

Turnbull and his coauthors suggest that rational bases for improvement are available in the form of limits—theoretical and engineering. Theoretical limits provide "both an outer bound for forecasts of potential process performance and a framework for clarifying the principles that govern the process." Based on fundamental principles and reasoning, they are numerical estimates of process performance. Engineering limits, on the other hand, "provide numerical estimates of the levels process variables could attain, using known technologies." The engineering limit for a specific indicator of process performance is intended as a practical estimate of what is achievable without regard to possible adverse effects on other indicators. Although theoretical limits are expected to be universally applicable (within some particular domain), engineering limits take into account the local context and circumstances of a specific production system. Therefore, engineering limits should move more closely to the theoretical limits with the introduction of newer production technologies or with change in the local context (such as organizational changes that promote cooperation between design and manufacturing groups).

Similarly, while continuous improvement is an important foundation of world-class manufacturing, it must be supported by appropriate mechanisms to measure improvement and to define the appropriate limits or goals to prevent excessive and wasteful investment. Knowledge of the theoretical limits can provide a benchmark for expectations of future improvement (see Foster, 1986).

Identifying Critical Variables

When the complexity is significant enough to suggest that the development of a model is necessary to understand the relationships, dependencies and interactions among the variables, it is also likely that one is not able to identify directly which of the variables exert the most control over the performance of the modeled system. In a model of a single-server queue (such as a machine tool with one operator) Krupka (in this volume) illustrates the significance of identifying and focusing on the critical aspects of the system. He draws attention to the sensitivity of the throughput to the variance of the service time and arrival rate of the modeled machine:

> Small decreases in the service rate (which effectively shift the system to a higher level of capacity utilization) lead to a large increase in throughput time . . . (and) an increase in variability, either in the arrival or in the service rate, leads to a large increase in throughput time. . . . The prescriptions for reducing throughput time (or manufacturing interval) are the same:

> reduce variability in the system and strive to increase the service rate (Krupka, in this volume, p. 170).

Therefore, when models appropriately represent the systems, they can be used to help identify the characteristics of the system that determine its control and thereby provide a basis for systematic improvement.

Strategic Planning and Management

Models are useful for considering operational questions that confront the manufacturing enterprise and to explore, in a timely way, strategic alternatives for the firm. Mize (in this volume, p. 196), comments on the changed context in which manufacturing managers find themselves today:

> Managers are rapidly losing many of the planning aids that have allowed them to proceed in an orderly, progressive fashion. In the past, managers could safely assume that tomorrow will be much like today, with only marginal changes. In fact, randomness was often much larger than the average marginal change; thus, the "noise" masked the "signal." Consequently, many of today's managers know how to manage only on the margin, in a static mode. Today's managers are faced with the fact that change is continuous, pervasive, and often traumatic. . . . A rapidly changing total environment has become the norm, replacing the relatively stable and static environment of the past.

Mize goes on to characterize the challenge of working backward from a desired future state to the present in a way that clearly shows a path of action. He suggests that models will be needed to help most people to deal with the interdependent variables and dynamic changes affecting the necessary day-to-day control to achieve their organization's strategic visions.

Basis for Decisions and Predicting Performance

Models provide a rational basis for predicting the impact of decisions before their implementation by (quantitatively) describing the important elements, interactions, and dependencies. Empirical models comprise valuable knowledge that provides a basis for engineering and managerial practice. Even simple models like the 15/85 rule described by Bowen (in this volume) are useful drivers of improvement and change.

The construction and continued refinement of models also make it easier to evaluate and transfer the assembled know-how from individuals and groups to others in the organization. Lardner suggests, that as a vehicle for capturing and conveying organizational knowledge, models are a "more accurate process than depending . . . on the experience of a few people and what they remember about the past."

Factories As Human Phenomena

A major difficulty with the topic of future factories is that the mind usually grasps it visually, as a static picture. But a snapshot view of the future factory is at best incomplete. It ignores continuing developments in technology, and it encourages debate about the desirability of specific renderings of technological possibilities, forms unlikely to appear in any event, far less to be influenced by the debate.

However, if we focus on the process of the design of future factories, a topic far more significant than any specific technological possibility, such as the robot, or for that matter any specific picture, such as the totally automated factory, three issues must be considered:

1. A factory does not appear suddenly in full operational maturity, but rather is continually designed and redesigned, implemented, constructed, and rebuilt.
2. It is not automatically programmed to improve. Continual energy and direction must be employed if it is to adapt successfully to changing needs and potential.
3. It will never be entirely free of people, never be completely automatic or robotic in this sense.

The factory is a human phenomenon. Every step from conception to eventual destruction is for, by, and because of people.

SOURCE: Nadler and Robinson (1983).

Pritsker (in this volume) also draws attention to the use of simulation in manufacturing companies as a mechanism for explaining and distributing complex rules and policies throughout the organization, especially to the operational areas on the factory floor. Using the same data to drive models throughout the enterprise, allows shop floor workers to acquire a perspective of operations that is in concert with the goals of the manufacturing system.

Models can be immensely powerful competitive weapons when used to capture particular competencies of the manufacturer and then leveraged throughout the enterprise that developed them. They offer an important means of accomplishing organizational learning as they extend their use well beyond a particular control activity.

Models should also be considered as a basis for evaluating continuous improvement efforts and changes made in the manufacturing system. Predictive capabilities of models are especially important when dealing with uncertainty about the nature of the problems being addressed and about the likely result of any proposed solution. Lardner (in this volume) emphasizes that the complexity, uncertainty, interdependence of the many elements of the manufacturing system, and the reliance on the experience of individuals are significant impediments to "good, timely decision making."

Simulation

Efforts to discover appropriate mathematical formulations for expressing and predicting performance are important for extending the science of manufacturing. However, much of the complexity and interdependency found in manufacturing systems does not readily lend itself to such rigorous and exact descriptions. A frequently used method for describing and exploring manufacturing systems is simulation:

> Manufacturing models analyzed by simulation (simulation models) are developed to study the dynamics of the manufacturing system. Such models are built without having to fit the manufacturing system into a preconceived model structure because the analysis is performed by playing out the logic and relationships included in the model. . . . Of fundamental importance is the building of simulation models iteratively allowing models to be embellished through simple and direct additions (Pritsker, in this volume, p. 205).

Manufacturing organizations offer a rich variety of opportunities for using simulation modeling. For example, it can be used to explain operating procedures, often through animations of the manufacturing system being modeled; to present graphical summaries of large volumes of data generated by the system, including tabulations, statistical estimators, statistical graphs, and sensitivity plots for analysis of manufacturing systems; to rank and select from among design alternatives; to schedule production; to dispatch resources; and to train machine operators, schedulers, and process design engineers.

> FOUNDATION: World-class manufacturers seek to describe and understand the interdependency of the many elements of the manufacturing system, to discover new relationships, to explore the consequences of alternative decisions, and to communicate unambiguously within the manufacturing organization and with its customers and suppliers. Models are an important tool to accomplish this goal.

5

Organizational Learning and Improving System Performance

Transferring philosophy is much harder than transferring technology.

—Donald F. Ephlin, Retired Vice President,
United Auto Workers

Corporations that compete successfully with the world's best producers will increasingly find that they have become members of a select group of enterprises that accept and practice the foundations described in this book. Performance at the level that is feasible with these foundations can be expected to become the norm for those enterprises striving to be the best-of-the-best. What then, it might be asked, will be the basis by which any of these firms will find it possible to obtain a competitive advantage over the others—all of which will be performing at a highly competitive level? From this plateau of competitive behavior, how can any firm hope to achieve a differential advantage over any other firm that is making effective use of these foundations?

There are two generally separate but related answers to these questions. The first is the ability of the organization to learn and improve. The organization that can learn more rapidly from its experiences and use that learning to enhance its performance will have a distinct advantage. The second answer is to be found in technology. The enterprise that develops the ability to lead in the effective use of technology will possess a distinct and important advantage over competitors. Both of these areas will be explored in detail in this chapter.

FOUNDATIONS RELATED TO LEARNING AND RENEWAL

The opportunity for manufacturing organizations to achieve and retain a competitive advantage depends to a significant extent on their ability to respond rapidly to changes imposed from the outside and to initiate changes aggressively. Two interpenetrating subsystems that are critical to transforming a firm are those related to learning and to organizational renewal. Mize (in this volume) notes that the environment in which manufacturers operate today is characterized by rapid and continuous change, contrasting with the relative stability of the past. In his view, the quality and effectiveness of the response to changing circumstances is based on the ability of the members of the organization to create realistic visions of the future.

A critical ability supporting this approach will be the systematic capture of the knowledge and wisdom gained by the organization. The successful organization must be capable of learning from its experiences and using that knowledge to respond to its ever-changing environment.

Besides learning from direct observation, the organization gains a further competitive advantage when it understands, and is able to describe formally, the capabilities and limits of its manufacturing processes. Turnbull and his coauthors (in this volume) suggest that an appreciation of past circumstances and performance and an awareness of the theoretical limits of the processes used provide "an upper limit for forecasts of potential process performance, and a framework for clarifying the principles that govern the process." Understanding the theoretical limits helps manufacturers to establish their priorities for advancing current capabilities as well as defining the threshold beyond which there is little opportunity for cost-effective improvement.

The Learning Process

One theory of learning holds that individual learning occurs when individuals detect a match or mismatch between outcome and expectation. If there is a mismatch, the individual moves to an error-correction mode, while a match reinforces the actions that led to the particular result. Learning is defined as the time when the individual discovers the source of the error and develops a strategy or means for correcting the error to return to the established norms. New strategies must be developed or invented on the basis of new assumptions to correct the error. Error correction, then, is "shorthand for a complex learning cycle" (Argyris and Schon, 1978).

The ability to remain stable in a changing environment is described by Bateson (1972) as single-loop learning. A single feedback loop maintains the level of performance of those organizationally established norms that can be expected to remain largely fixed even within a changing environment—for example, norms that relate to product quality, sales, or task performance.

At the organizational level, the learning cycle includes many of the same features—identification of norms that result from the rules and policies of the organization, attempts to discover what is necessary to modify current performance to achieve the desired state, analysis of the success of actions to accomplish change, and incorporation of the most successful actions into the operating fabric of the organization. For the feedback process to be effective, at both the personal and the organizational level, it is essential that clear measures—the appropriate metrics—of current performance must exist.

Welliver (in this volume, p. 235) identifies the importance of developing appropriate metrics (he calls them benchmarks) and communicating them in policies that members of the organization can use to identify problems at variance from those norms:

> A basic element in any TQC (total quality control) effort is communication of data—specifically, statistics and information that describe a problem or establish a benchmark for improvement. Awareness of problems is what maintains the sense of urgency among managers to initiate changes that lead to improvement.

Edmondson alludes to this organizational learning as he discusses "understanding what your customer wants." An understanding of the customer must extend across a broad range of functional departments in the manufacturing organization. The ability of the organization to integrate these wants with its own strengths, special capabilities, and expertise represents a form of organizational learning.

Single-loop learning is concerned with maintaining assumptions and maximizing effectiveness within the constant framework of norms for performance. In many situations, however, conflicts will arise between desired performance and existing norms and strategies. The response of the organization to these conflicts leads to what has been termed double-loop learning. The organization enters into double loop learning when the system begins to receive signals that the norms themselves need to be examined and perhaps modified (see Argyris, 1991).

Mize (in this volume, p. 200) captures the essence of single- and double-loop learning in the following way:

> The control system operates at two levels. First, it monitors a simulation of the future iteratively until an acceptable organizational strategy has been identified consistent with the vision of the desired future state. In a sense, this control structure is a feed-forward control loop.
>
> Second, a feedback control loop tracks actual results, compares them with the planned results emanating from the organization strategy, and determines appropriate corrective action relative to operational performance. It is important to note that this model captures corporate experience and

> imbeds the 'knowledge' accumulated from strategic and operational experience into the 'corporate memory' for use in future planning.

As we have seen, measures of key metrics must exist in order to assess the current level of performance and the magnitude of actions undertaken in response to the controls. Those metrics must be chosen in such a way that the system can measure its improvement against near-term objectives and determine realistic long-term goals. Addressing the ultimate levels of performance that the system seeks to achieve, Turnbull (in this volume, p. 229) argues that "For each process that is examined, potential actions are considered within the context of such questions as, Can improvement efforts make a significant impact, or is this process nearing its theoretical limit?"

The appropriate metrics can be used to define how fast learning must take place. Compton and coauthors (in this volume, p. 115) discuss the importance of improving quality through the concept of learning curves. They note that "the systematic collection of data on quality . . . [offers] a means of tracking progress on the 'continuous improvement' of quality and a means by which realistic expectations can be established for future goals. Above all, the existence of a learning curve for quality should be viewed as one more example of the need for careful collection of systematic data." The use of knowledge, derived from progress and experimentation, by all members of the organization represents an example of organizational learning. The rate at which an organization improves its performance as a result of learning is perhaps one of the principal determinants of whether it can become best-of-the-best.

The goal of the world-class manufacturer must be to make the information and knowledge available at the right place and time. Too often in manufacturing, sources of information become scattered and isolated. Welliver (in this volume) discusses a situation that is all too common on production lines—wide separation in both space and time between machining operations and subsequent conformance inspection activities. Operators producing parts do not learn about the quality of their work in a timely way. Information that was available to the operator machining the parts is lost once a batch of components leaves the area. The inspection, which was performed only at the completion of all production activities, was historically an industrial engineering function. When parts were rejected—new information was created—no linkage or feedback was established between the operator and the errors. In the new system, the operator collects data on each part as it is machined by measuring it against the criteria previously available to the production inspector.

Welliver also describes a situation in which an effort was made to transfer individual learning to the organization. Information about the performance of the machines was made available to operators on successive

Information-Based Organizations

The typical large business 20 years hence will have fewer than half the levels of management of its counterpart today, and no more than a third the managers. In its structure, and in its management problems and concerns, it will bear little resemblance to the typical manufacturing company, circa 1950, which our textbooks still consider the norm. Instead it is far more likely to resemble organizations that neither the practicing manager nor the management scholar pays much attention to today: the hospital, the university, the symphony orchestra. For like them, the typical business will be knowledge-based, an organization composed largely of specialists who direct and discipline their own performance through organized feedback from colleagues, customers, and headquarters. For this reason, it will be what I call an information-based organization.

The information-based organization requires far more specialists overall than the command-and-control companies we are accustomed to. Moreover, the specialists are found in operations, not at corporate headquarters. Indeed, the operating organization tends to become an organization of specialists of all kinds.

Information-based organizations need central operating work such as legal counsel, public relations, and labor relations as much as ever. But the need for service staffs—that is, for people without operating responsibilities who only advise, counsel, or coordinate—shrinks drastically. In its central management, the information-based organization needs few, if any, specialists.

Because of its flatter structure, the large, information-based organization will more closely resemble the businesses of a century ago than today's big companies. Back then, however, all the knowledge, such as it was, lay with the very top people. The rest were helpers or hands, who mostly did the same work and did as they were told. In the information-based organization, the knowledge will be primarily at the bottom, in the minds of the specialists who do different work and direct themselves. So today's typical organization in which knowledge tends to be concentrated in service staffs, perched rather insecurely between top management and the operating people, will likely be labeled a phase, an attempt to infuse knowledge from the top rather than obtain information from below.

SOURCE: Drucker (1988a).

production shifts by publicly displaying it on a large tablet or board. The sharing of information between shifts gradually improved the production process to the point that the variance of the process approached zero.

Individual learning experiences are not automatically converted to organizational memory and made available for all members to draw and build upon. The people with access to new data and information, authority to make changes, and understanding of the proper use of such information should devise appropriate methods to codify their knowledge so that it can be compared with information acquired by others. This requires that the organization's knowledge base be continually changed and updated. In some instances this knowledge may be specific to a machine or process, as Welliver's example shows.

Organizational learning is a broad-based strategy for capturing and making available to members of the organization information and knowledge that enable them to benefit from the experience of others—that build on the knowledge of many members of the manufacturing enterprise. In other cases, the knowledge of individuals and groups can be captured by tools, models, formula, drawings, instruction manuals, and the artifacts of the production process itself. Benchmarking the internal practices of the organization between projects is important.

Formalize Organizational Knowledge with Models

Members of an organization acquire information, analyze situations, react to stimuli, and reach conclusions concerning events. Models are important in capturing the critical variables and the relationships that have been discovered by members of the organization. Models, therefore, offer a broad basis for conveying shared experience and knowledge in manufacturing enterprises. Examples of simulation model applications and appropriate output types are presented by Pritsker (in this volume). These examples include the use of models as explanatory devices, such as animations of the physical production system; as analytic tools for statistical evaluations; and as educational devices when it is not possible or practical to use the system itself for hands-on training in a classroom (such as system failures or events that are difficult or dangerous).

Pritsker emphasizes the importance of developing models that can be used as a basis for long-term understanding of manufacturing processes and that contribute to the improvement of the manufacturing system:

> Models contain information about manufacturing processes and by using such models continually, the processes will be better understood. Understanding leads to improved manufacturing and information for improving design. Thus, TCM [total capacity management] is a mechanism to achieve, using simulation, a new form of Kaizen (Imai, 1986) by which the pro-

> cesses of manufacturing and decision making can be continually evaluated, changed, and improved. . . . Innovation also is enhanced, because a model developed in one functional area can be used to indicate the possibility of new constructs for another functional area. Thus, improvement cycles in a single functional area may be used to foster new models and concepts in other functional areas. The common model, common data foundation . . . when fully implemented, provides a basis for achieving world-class manufacturing (Pritsker, in this volume, p. 208).

World-class manufacturers recognize that they must be able to respond to externally induced change, but they must also be concerned with instigating continuous change within their enterprise. Organizational learning is a strategy for responding to environmentally induced change. The ease with which the organization incorporates the learning of individuals into its collective memory will be a major factor in becoming and remaining a world-class manufacturer. The time pressures afforded by global competition have greatly reduced the value of serial learning experiences. The world-class manufacturer must develop parallel programs for experimentation and creative destruction of the status quo with new technologies, machines, and techniques that will become the basis for the next generation of products and processes. The 15/85 rule—that 85 percent of project expenditures are determined during the first 15 percent of project time—should drive the organization not only to improve the introduction time of products and processes, but to also try and change the rule (Bowen, in this volume).

Although organizational learning depends on embedding into the organization's memory the discoveries and evaluations that have been found in practice, a means must be provided by which new practices are explored and perfected. The organization's support and encouragement for experimentation, its tolerance of errors that arise in pursuing improvement, and its attitude toward careful and open discussion of the causes of poor performance critically determine how successful it will be in learning and progressing. The managers of the business must demonstrate their commitment to evolution through change and associated risk by providing the necessary support and a consistent policy toward change that "reinforces the notion that it is good to take risks" (Marsing, in this volume, p. 195). An organization that regularly "shoots the messenger of bad news" will soon find that no one is willing to take the risk of trying new things, let alone carry the news about the experiments that are less than successful (additional discussion of organizational learning and renewal can be found in Senge, 1990, and Stata, 1989).

> FOUNDATION: World-class manufacturers recognize that stimulating and accommodating continuous change forces organizations to experiment and assess outcomes. They translate the knowledge

acquired in this way into a framework, such as a model, that leads to improved operational decision making while incorporating the learning process into the fundamental operating philosophy of the enterprise.

FOUNDATIONS THAT RELATE TO TECHNOLOGY

U.S.-based manufacturers have often adopted the view that technological prowess is a viable means of compensating for other shortcomings. The infrequent mention of technology in the above discussions reflects the committee's strong conviction that an enterprise can make the best use of technology only after it has embraced and is practicing the foundations described above. Only then can technology become a powerful force in achieving a competitive advantage.

Both the United States and its international trading partners stand to benefit technically and economically from the closing of the technology gap between nations and the increasing cross-fertilization in engineering, technology, and management (National Academy of Engineering, National Interests in an Age of Global Technology, 1991). For management, however, selection of the proper technologies from among technological opportunities is becoming a complex challenge that may be different for each manufacturer and for individual facilities. Choice of the appropriate technologies will increasingly consider local circumstances such as environment, work force, materials availability, relative costs of production, and the abilities of competitors (National Research Council, Toward a New Era in U.S. Manufacturing, 1986, pp. 32-33). By making use of leading-edge technologies, a manufacturer may be able to achieve lower costs, better quality, or greater customer satisfaction with existing products my making low-cost variations in small lots and thereby realize significant competitive advantages (see Report of the National Critical Technologies Panel, 1991). As Fisher points out (in this volume) the greatest potential for achieving such advantage exists in industries where the pace of technological change is slowest.

To accomplish these objectives, each enterprise must develop a strategy, both corporate and local, that encourages the search for the best and most important technologies, develops a procedure for effectively analyzing technological opportunities, creates or acquires the expertise needed to implement those technologies, and commits the necessary financial and human resources to introduce the new developments when they become available. Viewed from the perspective of the manufacturer, the technological opportunities are enormous. Opportunities for new technical initiatives abound in unit processes, material substitutions, the management of subsystems and

National Critical Technologies

The timely development and deployment of technologies is essential to satisfy such national needs as defense, economic competitiveness, public health, and energy independence. Identification of technologies for concentration of effort becomes, therefore, a matter of considerable importance.

The National Critical Technologies Panel, appointed by the Director, Office of Science and Technology Policy, Executive Office of the President, identified a set of technologies that reflects the full range of national technology needs. Nearly 100 separate technologies were nominated by the Panel for consideration. Based on selection criteria and extensive private sector and government input, the Panel selected the 22 they considered the most important.

Materials

Materials synthesis and processing
Electronic and photonic materials
Ceramics
Composites
High-performance metals and alloys

Manufacturing

Flexible computer integrated manufacturing
Intelligent processing equipment
Micro- and nanofabrication
Systems management technologies

Information and Communications

Software
Microelectronics and optoelectronics
High-performance computing and networking
High-definition imaging and displays
Sensors and signal processing
Data storage and peripherals
Computer simulation and modeling

Biotechnology and Life Sciences

Applied molecular biology
Medical technology

Aeronautics and Surface Transportation

Aeronautics
Surface transportation technologies

Energy and Environment

Energy technologies
Pollution minimization, remediation, and waste management

SOURCE: Report of the National Critical Technologies Panel, March 1991.

interfaces between subsystems, and the description and control of system performance and response to changes. In each of these areas, as well as many others, research focuses on improving the effectiveness and efficiency of one or more elements of the manufacturing system. The challenge to the manufacturer is to develop a capability to access and harness this burgeoning research activity.

Since the technologies involved in the unit processes are often quite different from those involved in the total system, it is appropriate to divide the following discussion into three parts: the unit processes and subsystems used in material transformations, the interfaces between these many subunits, and the manufacturing system as a whole.

Unit Processes and Subsystems

The unit processes in manufacturing encompass a vast array of materials, material transformations, operations that combine and join materials, and assembly, testing, and inspection. Much of the research on unit processes is focused in university research laboratories. A theoretical understanding of the underlying processes must be based in the laws of physics, chemistry, metallurgy, and fluid dynamics and is frequently the subject of study. In many cases such study leads to improved materials, better process controls, and enlarged capabilities for applications. Suppliers of manufacturing subsystems are also active in research on processes relevant to the products they market. Turnbull and coauthors observe (in this volume, p. 226) that for complex systems, understanding of processes occurs at various levels:

> At the top level, it is valuable to have a balanced, descriptive understanding of the process, including measures of efficiencies, quantities of output, and the attributes of the output. The most profound process understanding, however, requires an examination of the underlying principles, mechanisms, and root causes.

Rather than attempting to summarize or evaluate the array of research activities under way on these topics, we will address the issue of access and effective use of research results by the manufacturer. There is no lack of information about the content of research activities related to unit processes and subsystems. The technical literature, the meetings of the professional societies, and the trade journals are all valuable sources of such information. Often the results of research are presented as isolated studies of rather detailed phenomena. The challenge confronting the manufacturer is timely evaluation of the potential impact that any given project may have on the overall operations of the enterprise. A manufacturer not only must be knowledgeable about worldwide research activities but must also possess

Recognizing Technological Limits

In any field, technological improvement is eventually limited by the laws of nature. The ultimate strength of a fiber is limited by the strength of its intermolecular bonds. The number of transistors that can be placed on a silicon chip is limited by the crystal structure of the silicon material. The goal of technical management is to identify the limits of any given technology early, as a first step in determining what finally can be accomplished with it.

Industry, though, is usually far from these natural limits, and it is more likely to come up against practical, physical barriers that represent the current state of the art. The difference between the technical and the state-of-the-art limits determines the technology's potential for performance improvement; the greater the distance, the more the potential. This can have dramatic strategic implications.

If the concept of technical limits is well understood and thoughtfully acted on, the task of planning for adoption of new or alternative technologies is relatively straightforward. Through an informal process of estimating the company's proximity to its technological limits for currently employed technologies, managers can begin to assess their next moves. This is best done analytically, although it can be started as an intuitive process. [Some] key signals that suggest trouble when the company approaches the limits of an existing technology:

- An intuitive sense among top managers that the company's R&D productivity is declining.
- A trend toward missed R&D deadlines.
- A trend toward process rather than product improvement.
- A perceived loss of R&D creativity, and disharmony among the R&D staff.
- Lack of improvement from replacement of R&D leaders or staff.
- Profits that come from increasingly narrower market segments.
- Loss of market share—particularly to a smaller competitor—in a specialized market niche.
- Little difference in returns despite spending substantially more—or less—than competitors over a period of years.

Assessing each of these points is likely to spark a rigorous investigation of the company's technologies. It may prompt the company to consider alternatives, and management may discover previously unnoticed discontinuities and the potential for future transitions. One thing is certain: Assumptions grounded in evolutionary, incremental thinking will be severely tested.

SOURCE: Foster (1982).

insight into the effect of integrating a particular improvement into the system. (See Kelly and Brooks, 1991, for discussions on adoption of new technologies in manufacturing firms.)

Subsystem Interfaces

Although the unit processes and the subsystems of a manufacturing enterprise are the focus of a great deal of attention, the interfaces between them frequently receive less attention from the research community. This is perhaps understandable since a study of the interfaces demands that one treat the complexity of the total system:

> One central activity, introduction of a new product or process, is itself a system with all the risks and uncertainties of complex systems. When considering the enterprise or any element of it, understanding becomes richer when one looks for the interconnections of activities, functions, processes, and outcomes (Bowen, in this volume, p. 99).

The importance of the interaction between subsystems cannot be overemphasized. Information must flow throughout the organization, thus crossing the subsystem boundaries, and decisions affecting subsystems can affect overall system performance. Management of the subsystem interfaces often represents a substantially greater challenge than management of unit processes, because improvements at one stage in an integrated production process can throw a downstream processing step out of control. As Marsing points out (in this volume), understanding the interdependencies between processing steps is critical in minimizing risk associated with making changes in an integrated system.

Issues that must be confronted include those associated with simultaneous engineering—with the creation of a process that encourages each of the various groups in an organization to participate and contribute to the design of the products they will be expected to make, sell, and service. In the typical organizational structure, simultaneous engineering requires the creation of groups whose members represent the functional divisions in the company. Cook (in this volume, p. 125) points out that the cultural differences between these divisions can make their interaction difficult. The "throw-it-over-the-wall" syndrome for product realization most likely arose from the desire to minimize face-to-face interactions between functional divisions after transactions had grown to be too tedious and adversarial as cultural differences became large and entrenched over time. Cook says, "The sharp differences in operational responsibility between divisions in the functional organization are most likely the root cause for their sharp cultural differences."

Two approaches can be taken to solving these problems: Cook suggests that a different organizational structure is needed to reduce the problems. Others are working to develop tools and procedures that will reduce the barriers within the typical functional organization. Some of these efforts relate to improving the effectiveness of group participation, including development of such computer-based aids as expert systems or artificial intelligence systems. Other efforts focus on the needs of the subgroups as information is moved throughout the organization.

Technology and Manufacturing as a Continuum

Just as science and technology are a continuum, so are technology and manufacturing. Technology is the driving force in the design and manufacturing of products and the development of services. Technological innovation, design, and manufacturing are interactive and interdependent processes. For example, product development, design, and manufacturing all drive research and are, at the same time, highly dependent on research for successful innovation.

In contrast to its competitors—particularly those in Japan—U.S. industry has too often placed artificial boundries between research and design on the one hand, and production and marketing on the other. This counterproductive philosophy is embodied in many American companies in the physical separation of research facilities from the factory floor. Today's trends in manufacturing are toward shorter production runs, shorter product cycles, increasing quality, a greater variety of products with increasing customization, and a shorter time to market for new products. This new climate demands an integrated manufacturing environment that facilitates making incremental improvements and refinements in both manufactured products and the manufacturing process.

There is consensus that the United States does not lag behind its competitors in generating basic research results or in the quality of its doctoral scientists and engineers. But America's ability to rapidly translate science and engineering into commercializable products, and its ability to take advantage of the detailed insights, understand, and processes that are a prerequisite to product design and product manufacturing is another story altogether. Thus, in recent years it has become apparent that a "manufacturing gap"—like the technology gap of earlier years—has emerged, and this time it is the United States that lags behind.

SOURCE: Bloch (1991).

Lardner notes (in this volume, p. 176) that

> The disappearance of a common language among the many groups in manufacturing highlighted a previously unappreciated problem in data and information management. This is the task of translating data and information from the root sources into the format and language needed by functional groups without losing the precise intent and meaning of the original. It is apparent that there is a serious lack of discipline in the data and information management systems of most manufacturing companies and that this lack of discipline perpetuates.

Hanson (in this volume) insists that a distinguishing characteristic of the Integrated Enterprise is the comprehensive communications network that facilitates the open distribution of knowledge and information. And Lardner notes that inadequate information systems have contributed to breaking the manufacturing whole into many incongruous parts.

Research related to the subsystem interfaces tends to be more difficult than research on the unit processes and the subsystems themselves, largely because of the increased complexity of the problem. It is difficult to develop a laboratory representation of the interfaces and more difficult still to represent the complexity of the total problem. While a great deal of effort has been devoted to developing tools for the effective planning and scheduling of machines, materials, and people—all of which recognize the existence of the interfaces and attempt to deal with them—a large segment of the research tends to focus on questions that are more generic. Making use of the results of this latter type of research for a particular manufacturing environment requires, therefore, the involvement of people who are intimately involved with both aspects of the problem—the research issues and the practical problems encountered in the manufacturing environment.

System Issues

Lardner (in this volume, p. 177) describes the issues affecting technological advances at the system level in the following way:

> Since objectives and values vary from group to group, and there is little or no understanding of how the actions of one group will affect all the other groups, responses to changes in the manufacturing environment vary greatly. It is almost by accident that group actions are directed toward optimization of the whole manufacturing effort.

This does not mean, however, that there are no tools that are useful at the system level. As expressed by Herbert A. Simon (1990; quoted by Pritsker in this volume, p. 205):

> Modeling is a principal—perhaps the primary—tool for studying the behavior of large complex systems. . . . When we model systems, we are

> usually (not always) interested in their dynamic behavior. Typically, we place our model at some initial point in phase space and watch it mark out a path through the future.

The growth in modeling and simulation of manufacturing systems in the past decade follows from recognition of the need to improve manufacturing operations and the need to assess the effect of decisions before they are implemented. The availability of simulation languages to build and analyze manufacturing models has stimulated this growth.

Research in these areas is extensive and broad based, and new tools constantly being created through the work of both universities and private concerns. The challenge to the researcher is to create the tools in such a form that they can be easily used. Solberg (in this volume) proposes that the power of a model or a modeling technique depends on its "validity, credibility, and generality" and that the simplest model that expresses a valid relation is usually the most powerful.

Enhancing the Scientific Method for Understanding Manufacturing Systems

For years the scientific method has been presented as a way of understanding natural and man-made systems by constructing hypotheses and testing their validity either in a laboratory or in the real world. The development of computer modeling and visualization graphics methods has made it possible to enhance the scientific method as applied to manufacturing systems. The increased ability to model in both mathematical and logical terms, the advanced computational analysis procedures available on modern computing facilities, and the large improvements in the display of both static and dynamic data on graphics terminals, are illustrated by the three steps added to the scientific method in Figure 2.

In manufacturing systems, the development of theories is difficult. For this reason, models of manufacturing systems contain many conditional relations and not a large number of mathematical equations. In many cases, manufacturing data are used directly in the model although sometimes a mathematical characterization of the data is employed. For mathematical models of manufacturing systems, the analytic technique most often used is simulation. The outputs of a simulation analysis take the form of plots of variables over time. For example, the status of machines, fixtures, and tools is shown as a percentage of time in various states, such as in-use, available, being maintained, or broken. The visualization of status is typically shown in either a pie chart or a bar chart. Manufacturing throughput is presented as a number of finished products produced over time, and the time required to manufacture a finished product is usually presented as a histogram of production time. In addition, statistical estimators of the simula-

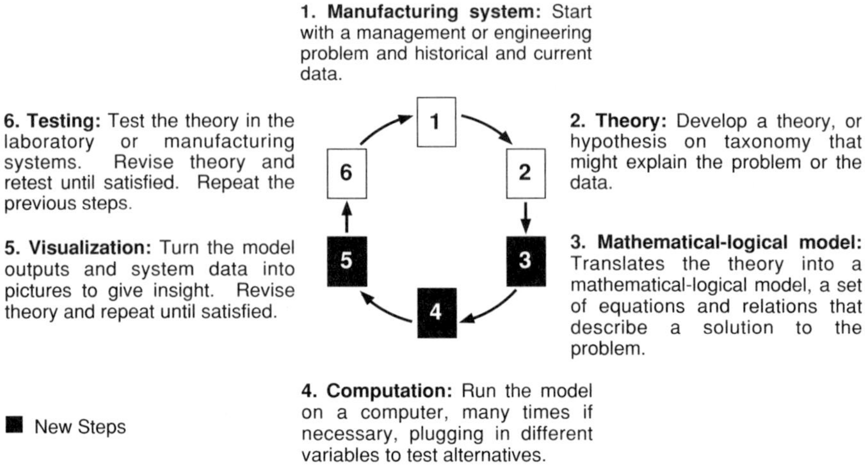

FIGURE 2 Enhancing the scientific method with advanced computational analysis and improvements in the display of both static and dynamic information. (Adapted from Cornell Engineering College Information as presented in the Indianapolis Star, January 20, 1991.)

tion outputs are developed, and measures that relate to direct manufacturing variables such as cost, adherence to schedule, and quality are calculated. The complete operation of a model of a manufacturing system is best visualized in an animation in which icons are used to represent resources, facilities, and parts. The movement of the part icons through resource icons portrays the dynamic operation of the manufacturing system.

Currently, modeling and simulation are used to test alternative ways of operating a manufacturing system to satisfy the objectives of the corporation. In the future, relationships between output performance measures and input data will be sought. These will provide the basis for theories of manufacturing system operation. By performing many iterations of the scientific method cycle shown in Figure 2, theories of manufacturing systems operations will be developed.

The challenge to the management of manufacturing enterprises is to insist that the available tools be regularly and consistently used to analyze the systems. This, in turn, means having people available who are either trained in the use of these tools or in the development of a relationship with other organizations that will do the analysis for them. In either circumstance, the important issue is to insist on the use of the best and most effective tools for analysis.

Just as Krupka (in this volume) has argued that time is a critical metric in evaluating the performance of a manufacturing system, so too is timeliness a critical element in ensuring that new research results produce a competitive advantage. The earlier a result is known, the earlier its potential can be analyzed and a plan developed for its application. This argues strongly for the development of a strategy that will couple the user, in this case the manufacturer, with the research activity, whether it is in the university, the host company's research laboratory, or a supplier's laboratory. Frequent communication and visits, exchange of people, and joint projects are just a few of the mechanisms being developed to enhance the early understanding of research results and to enhance the capability to support and guide the direction of the research activities. As noted above, to be successful, this requires the commitment of personnel and resources to the task. To be successful, these efforts must be viewed as a long-term investment in the competitive posture of a company.

> FOUNDATION: World-class manufacturers view technology as a strategic tool for achieving world-class competitiveness by all elements of the manufacturing organization. High priority is placed on the discovery, development, and timely implementation of the most relevant technology and the identification and support of people who can communicate and implement the results of research.

6

Educational and Technological Challenges

> [Today] the changes that so many companies are making are more than a response to "globalization." They denote nothing less than the obsolence of the corporate model many of us have grown up with. For some people, it won't be easy to let go of old concepts, old hierarchies, old sources of power—but it's mandatory to think anew.
>
> —Vernon R. Loucks, Jr., Chairman and CEO, Baxter International, in Review, 1990.

It is the committee's strongly held conviction that the worldwide competitive environment will richly reward the manufacturers who adopt the foundations of manufacturing while penalizing those who do not. This report has focused principally on the foundations and the rationale for identifying these as keys to achieving world-class status. It concludes with a few brief comments concerning the actions that must be taken to make the foundations common practices and some suggestions concerning the likely implications of their adoption.

The challenge posed by the worldwide competitive environment will demand many things of the enterprise, particularly leaders who understand the system of manufacturing, its elements and their interrelationships. Leaders of successful enterprises will have an enhanced understanding of the capabilities of their competitors; they will strive to eliminate organizational parochialism; and they will develop the ability to respond more quickly to market changes. Realizing the benefits of a positive response to these challenges will place unusual demands on all of the participants of the enterprise.

Although success in implementing the foundations depends on many things, the committee emphasizes that they represent a system of actions that cannot be embraced piecemeal. They are as interrelated and as overlapping as are the elements of the manufacturing system that they seek to improve. They must be viewed as a system of action-oriented principles whose collective application can produce important improvements in the manufacturing enterprise.

Effective communications are obviously a key element of success in implementing the foundations. But it must be recognized that it is easier to develop the mechanisms for collecting inputs from many sources than it is to create an environment that generates rapid, sensitive, and consistent responses to those many voices.

This is especially true for those broadly defined as customers. It makes little difference whether the customers are internal to the enterprise, such as those employed in the firm's production activities chain, or organizationally separate from the enterprise as purchasers and users of the products manufactured. Encouraging the active participation of customers in the affairs of the enterprise places important obligations on both the customers and the enterprise. Organizing the information, assigning responsibility for response, allocating resources, and training people to be good listeners is a daunting task for the enterprise. The customers must also learn the areas in which they have a rightful role in helping change and in ensuring that their comments are factual and responsible. Patience is required in developing effective communication channels and in identifying the proper issues to be discussed.

Just as the customer must become an increasingly influential member of the manufacturing system, so must also the suppliers be viewed as more than just a provider of materials or components. The long-term commitment of the enterprise to a few key suppliers and of those suppliers to the enterprise demands a level of understanding and trust that is not easily created, but once created must be constantly nurtured. This is likely to be a fragile relationship unless all parties recognize and are willing to work diligently to maintain it. Working as a family, sharing both the burdens and the rewards, and offering and accepting friendly and constructive criticism are not only honorable objectives but increasingly important factors in the success of the manufacturing enterprise in this competitive environment. To succeed demands a level of commitment and openness that may require a fundamental change in outlook on the part of everyone involved with and in the enterprise.

The creation of an environment that encourages employee involvement and employee empowerment as a means of achieving continuous improvement has been demonstrated to require an unusual commitment of time, resources, and training. The effort must be maintained and nurtured over a long period of time. A manufacturer is unlikely to succeed if these tasks are approached as interim measures in response to temporary market pressures. The successful manufacturing system will most likely be one that has evolved into an organization in which the participants work in harmony with an appreciation for the particular contributions of each member. Management, employees, suppliers, and customers will share common interests and goals and will create a team to work toward achieving those goals. It is the responsibility of the management to provide consistent leadership for the organization as it strives to reach these levels of integration. Wilson's jazz ensemble provides a model of the organizing, coordinating, and direction-setting determinants for successful manufacturing systems.

CHALLENGES FOR THE EDUCATION SYSTEM

The committee believes that the implications of these foundations are important to elements of our society beyond that normally described as the manufacturing enterprise. These include the educational system, which is training the next generation of employees, and the technical community, which is responsible for creating new understanding of phenomena and providing new tools for solving both current and future problems. Since neither of these sectors has been extensively discussed earlier in this report, the implications for them will be examined somewhat more fully in this chapter.

The committee has argued throughout this volume that the modern manufacturing enterprise cannot be competitive if it continues to operate as a loosely coalesced group of independent elements—elements whose identity depends on a discipline or a detailed job description. Gibson (in this volume, p. 150) emphasizes that "the elements of Taylorism are . . . no longer right for modern America." Gibson argues that for the educational system, the approach established by Taylor became "the universal paradigm" for engineering education, and, although we are in the process of changing that approach, Taylorism remains the predominate model for education today. In Gibson's view, "This academic process is patterned after the old Tayloristic suboptimization of individual operations on a manufacturing line with no thought for overall production efficiency." If, indeed, this is a proper description of the current focus in education, one must ask, as Gibson does, whether we are encouraging the proper outlook and training for those who will manage and operate the next generation of manufacturing enterprise. Gibson has called for a reassessment of the current approach to education for the next generation of practitioners or professionals, whether those professionals find themselves in management, technical, or nontechnical roles. If the success of the manufacturing enterprise depends on eliminating organizational barriers and enhancing communications, should we not expect that similar changes would be beneficial for the educational system that is training the people who will lead and operate these systems? Accomplishing changes of this extent in the educational system will require leadership with the same dedication and vision that has characterized the industrial leaders who are rejuvenating the industrial system. Gibson has suggested an in-depth analysis of what the rapidly changing industrial environment implies for the educational system.

THE TECHNOLOGICAL CHALLENGES

In Chapter 5 we saw that the successful enterprise will eventually find that technology will become a key ingredient in achieving a competitive advantage. The process by which needs are assessed and then both human and financial resources are committed to the search for new technologies is complex. In some cases this process will involve the support of substantial in-house activities that have the freedom to explore the far reaches of technology while remaining in close contact with the near-term problems of the enterprise. In other cases it may mean developing relationships with other companies, universities, or not-for-profit organizations. The proper mix of surveys with the development of technology is, and will continue to be, an important decision for management. Orchestrating these efforts and guiding them with vision and imagination will place special demands on management. It will, above all, demand a renewed understanding of, and an appreciation for, technology—qualities that have frequently been missing in the leaders of the nation's industrial complex.

As difficult as these tasks may be for the large organization, it may be almost impossible for some small organizations whose raison d'être is not technology, to use technology as a competitive tool. Many small companies have no research capability, and their resources are sufficiently limited that they are unable to allow their employees to focus on such matters. These small firms are often dependent on vendors and suppliers to bring them new ideas and opportunities. Some may obtain assistance from the principal customers for their products if those customers happen to be large firms and if the relationships are sufficiently mature to allow an unencumbered exchange of proprietary technical information. Others may be able to use the resources that state governments have introduced to encourage economic development, such as the state technical assistance programs, programs that involve information transfer through centralized data bases, and the research capabilities of local universities and technical colleges. Recognizing that about 40 percent of the value added in manufacturing is generated by companies with fewer than 100 employees, it seems clear that ways must be found to offer these small firms timely access to technological change.

Since many of these small manufacturers supply the larger enterprises, the competitiveness of the entire sector is critically dependent on the ability of each segment to improve. Whatever the source of help, it is clear that small enterprises, which constitute a significant fraction of the manufacturing capability of this nation, deserve considerable attention and assistance.

Ensuring that the technological infrastructure of the manufacturing sector—viewed in the national sense—is being supported at the proper levels presents a particular challenge to all U.S. manufacturers. With a largely decentralized system of supporting, managing, and disseminating information relative to technological developments, it is often difficult for any single organization to develop the breadth of vision needed to ensure that its own interests are being properly accommodated.

The implications of the competitive environment that has evolved over the past 20 years are profound. Just as no single element in the manufacturing system can ensure that an enterprise will be successful, so can no single sector of the national infrastructure ensure that the industrial sector will be competitive. A commitment to renewal of the U.S. manufacturing sector is essential. A willingness to learn from each other is critical. No one can afford to take the risk of waiting for others to show the way. All manufacturers must embrace the doctrine that continuous improvement demands their immediate and unrelenting attention. U.S. manufacturers cannot allow their competitors to set the standards by which success will be achieved and to be the leaders in meeting those standards. The United States must establish as a national goal a strategy that encourages and supports the adoption of the foundations of world-class manufacturing systems.

GLOBALLY COMPETITIVE MANUFACTURING PRACTICES

The authors of the 19 papers in this section draw upon their experience, knowledge, and expertise to explore many of the principles and practices that characterize world-class manufacturers. Differences in terminology and clarity among the papers are evidence of the communication barriers resulting from the functional specialties in manufacturing and the absence of a common manufacturing language. Despite the diversity of backgrounds and, therefore, the different perspectives from which they view manufacturing—as executives, university educators, and researchers—the authors express a consistent set of themes and concerns throughout this collection of papers:

- A "system view" is critical to understanding the key relationships, interactions, and interdependencies of the people and components needed to develop, produce, and market a firm's products.
- Manufacturing must move from a historically experiential basis and begin to develop the rigorous theories and foundations needed to understand, measure, control, and predict their performance.

- Management has the responsibility to involve and empower the work force to achieve the goals of the organization and must look beyond the walls of the factory to interact with customers, suppliers, and the educational community.

The breadth of the material presented in these papers effectively illustrates the scope of the challenge confronting manufacturers. The practices and concepts discussed here are central to an understanding of the concerns that U.S. manufacturers must address to become competitive in an expanding global marketplace.

Involvement and Empowerment: The Modern Paradigm for Management Success

NANCY L. BADORE

The history of industrial development is replete with challenges that have changed the direction of individual companies as well as entire industries. Firms and institutions that were operating successfully were forced to find imaginative and innovative new ways of operating or face the prospect of extinction. New ways or new technologies were needed to accomplish what had previously been thought impossible. The past decade—or perhaps decade and a half—will undoubtedly be noted as one of these critical periods in industrial history. Industry groups and firms that had dominated the scene since World War II were suddenly confronted with a new and very different challenge—worldwide competition. Reasonably inexpensive transportation and highly reliable communication systems, combined with new approaches to managing and controlling the manufacturing enterprise, enabled manufacturers in all segments of the globe to compete in markets that had previously been reserved to those who manufactured where the product was marketed. Furthermore, the plateau on which the competitive battle was joined focused not just on costs but on quality, responsiveness, and flexibility—all in the name of satisfying a reawakened interest in providing customers with what they needed or demanded.

While no unique set of elements properly describes all the companies or industries that were challenged by this wave of world competitiveness, many firms found that they suffered from some of the following characteristics. The customer was not recognized as having the determining influence on product attributes or performance; tensions existed among various units and between various levels of the enterprise; no mechanism existed for setting priorities among the many desirable corporate objectives; suppliers were treated as a necessary evil to be tolerated but not trusted; and management and labor were confrontational in attitudes and objectives.

Firms that manifested these characteristics, even though they were common industry practice, found themselves at a distinct disadvantage to competitors who emphasized the customer, provided a product of high quality, and maintained an internal working environment that was stimulating and cooperative. It is not an overstatement to say that many U.S. firms were confronted with the highly unpleasant alternative of changing their ways of doing business or facing extinction. It was in this environment—the new competitive environment of the mid-1970s to early 1980s—that U.S. firms turned their attention to the development of employee involvement.

WHAT IS EMPLOYEE INVOLVEMENT?

The term *employee involvement* means inclusion of the employee in the operation of the system. But it is the creation of the process—the tools and the means—by which this is accomplished that is the topic of this discussion. The two principal objectives of employee involvement are as follows:

1. To create, share, and make "real" for all employees a vision of the goals for the overall enterprise as well as for each organizational unit.
2. To seek and share the knowledge possessed by individual employees in achieving that vision.

This cannot be viewed as a grand philosophic statement of principle. Rather it must be an operational statement of a process that calls forth a new level of participation by the employees. This statement must serve as a guide by which each employee can support the shared goals of the enterprise while at the same time it serves as a standard against which the appropriateness of alternative actions can be assessed.

RELATIONSHIP OF EMPLOYEE INVOLVEMENT AND EMPOWERMENT

Employee involvement recognizes that individual employees have the best opportunity to understand and appreciate the problems that are unique to their positions; that the employees also have the greatest insight and experience in suggesting ways of solving those problems. It does not directly follow, however, that a mechanism is available by which that knowledge and experience can be put to use in solving those problems. This is the function of "employee empowerment." If proper advantage is to be taken of the knowledge that the employee possesses, it is necessary for an organization to empower the employee to implement the improvements that they know to be necessary. By so doing, the enterprise is making the employee an integral part of the process of staying competitive.

WHY WAS "EMPLOYEE INVOLVEMENT AND EMPOWERMENT" DEEMED SO CRITICAL?

Although size of the enterprise is not a prerequisite for being successful, certain aspects of "smallness" clearly make some tasks easier to achieve. Among these are the ability to achieve good communications among all employees, a common understanding of corporate and organizational objectives and goals, and a greater chance that the managers know personally the employees who must be involved in problem solving. The role of employee involvement and empowerment is, in a sense, the means by which a large organization attempts to achieve many of the benefits that are generic to the small organization. Although certain organizational structures and systems are required in larger organizations, the effort to accomplish meaningful employee involvement and empowerment is directed at preventing the organizational structure and systems from providing barriers to finding the best solutions to problems. Furthermore, the involvement process provides a means of humanizing the organization and maintaining participation by individuals at all levels—a process that is intended to lift the organization to new heights of competitive performance through the best use of the skills and interests of the individual. It is the means by which continuous improvement can be made an operating goal for all levels of an organization.

One might question, of course, the nature of the stimulus for developing employee involvement and empowerment. It would be pleasing to assert that recognizing the importance of employee involvement and empowerment was a natural part of organizational maturation and that the organizations saw it as a natural way to improve their performance. In fact, this was not the case and it was embraced, sometimes reluctantly, often belatedly, when the competitive crisis became so stark that management and workers alike were willing to take what was viewed as drastic action in order to survive. Industry appears to be no more immune to crisis management than any other segment of our society.

The process by which firms and enterprises achieved these changes has not been easy. It has been a difficult transition from the hierarchical structure, dominated by managers who "know what is best" for the organization and who expect others to do whatever task is asked of them, to a more open and sensitive environment in which people are willing to listen and respond to what they hear.

ACCOMPLISHING THE CHANGE

In discussing some of the issues and barriers that exist in accomplishing the transition to a more "people-oriented" company, I will draw on my experiences at Ford Motor Company. For simplicity, these experiences are presented as occurring in three distinct phases, but it should be realized that no such simple demarcation actually existed. In fact, some element of each phase was under way at all times.

I will try to offer some insight into the problems we experienced and the lessons we learned. The process of change is never easy for an organization. For a large, complex organization, it can be particularly traumatic. A retrospective assessment of the process and its results—after many of the frustrations have diminished in importance with the passage of time—can only be viewed as immensely valuable to the organization and to the people who have participated in it.

Phase I—Employee Involvement: Change at the Plant Level (1979-1982)

One should recall that the late 1970s was a period of economic downturn, in part a result of OPEC actions. It was also a period of severe labor stress with frequent confrontation between labor and management. This was the environment that provided the stimulus for experimentation and change. The change sought was originally cultural—improvement of the relationship between union and management within the plants. The vehicle to achieve the change was to be the establishment of quality circles, a somewhat unoriginal concept at the time, given that the quality circle movement was being imported widely from Japanese manufacturers.

What differentiated the Ford experience from other quality circle efforts at the same time—efforts that consistently ran out of steam—was the fortuitous combination of several elements. These included timing—the near-depression that hit the manufacturing sector at the same time the project was launched; the care taken to prepare each plant, one at a time, to install quality circles; and the vitality that was inadvertently built into the process when age-old adversaries—union leaders and line managers—were "married" to provide joint oversight to the change efforts. Perhaps the most critical element of all was the fact that quality, selected as the overarching goal for these joint efforts largely because it was noncontroversial, would turn out to have important consequences for the competitiveness of the company and the union alike.

In 1979 Ford's U.S. manufacturing organization consisted of some 80 plants and parts depots varying in size from several hundred to several thousand employees. Organizing to manage change on this scale meant applying concepts devised by researchers and thinkers who dealt with much smaller organizations. We could find no precedent for a change effort as extensive as the one we contemplated.

In this first phase of change at Ford, a formal organizational development model was created centrally to be implemented in a decentralized fashion—that is, on a plant-by-plant basis. This model drew heavily on the theoretical work of Walton (1969) at Harvard and of Beckhard and Harris (1987) and Schein (1969) at Massachusetts Institute of Technology. It drew also on our analyses of why so many of the attempts to install quality circles in the United States were failing in the early 1980s. The model presented, almost simplistically, a series of five steps to be taken by the leadership of an organization before embarking on a significant change, that is, before launching quality circles or "Employee Involvement Groups." We defined "leadership" as the senior-most positions in the plant on the company and union sides. That is, the leadership of the employee involvement efforts was to be provided only by a pairing of the plant manager and the union leader.

Creating a model, however theoretically valid and practical in its intent, was one thing; getting people to use it was another. "Marketing" the concept of employee involvement—and the business and political rationale underlying it—was a crucial part of the change process. The job of speaking on behalf of the change effort was undertaken by two people with high visibility and credibility in the plants: Peter Pestillo, Ford's vice president for labor relations, and Don Ephlin, vice president of the United Automobile Workers' Ford Department. For several months, on request, they jointly visited plants to hold informal meetings with line managers and union officials. They accomplished two things at these events: they modeled an unprecedented camaraderie and they educated their audiences about the link between business outcomes and the ways business is conducted.

At the end of 1979, several plants had volunteered to sponsor the first pilot employee involvement projects. These living laboratories were to teach the rest of the company about the real issues underlying a major change effort. We found, for example, that it was not difficult to enlist volunteers from management to improve quality or from the union's political hierarchy; nor was it difficult to enlist volunteers from the ranks of the people working in the plant. What *was* difficult was handling the sheer workload associated with implementing improvement ideas once the problem-solving groups began to meet and work.

Only belatedly did we discover that quality circles reverse the work flow of the traditional hierarchical organization: ideas for implementations were suddenly flowing upward from the broad base of our plant pyramids. There is much talk about management sincerity as the crucial factor in efforts to change quality. We were to learn that it is even more a question of being organized to handle the incremental work flow associated with putting new ideas to work: reviewing group recommendations and assigning priorities—including capital allocations when necessary—and responsibilities for implementation. Communicating back to the Employee Involvement Groups the final determination on suggestions, the status and timing of projects stemming from the accepted suggestions, and why certain suggestions would not fly, also took a tremendous amount of coordination and (for us) new sensitivities and skills.

Nonetheless, these early projects prevailed; they achieved remarkable results in improved quality and various organizational (and union) measures of payoff. By the end of the first three years, quality had been improved up to 40 percent by external industry measures. The projects, moreover, have persisted and have expanded and spawned many other significant initiatives.

It might be instructive to enumerate some of the lessons we learned from this early experience:

1. A business focus is essential in the early stages of a process this size—in our case the focus was on quality.
2. The goal must be viewed as worthy of achievement by all involved—it was this worthiness, not loyalty to management or union, that earned the time and commitment from the initial volunteers.
3. Every level of the organization has a crucial role—from the chairperson to the average worker on the factory floor. Leaving levels out can subvert the process.
4. "Early experimenters" (the first volunteers) in these change efforts are crucial to success. It will be their results, not the concept itself, that will attract the interest and support of a critical mass of the organization.
5. The organization must have the capability to learn from its early successes and failures and to transfer this knowledge to other elements of the organization.
6. Such efforts are sustained when they are tied to competitive improvement—ours has lasted through changes in union leadership, plant managers, vice presidents, and business cycles. "Cultural change for the sake of change" would not have lasted this long.

Phase II—Participative Management: The Change in Middle Management (1982–1985)

The early experience at the plant level demonstrated a number of things. First, it was clear that an immense reservoir of good ideas resided with the employees. Second, it was necessary to modify significantly the existing relationships among management and labor if we were to be successful in tapping this resource. Third, the process of change that was being pursued could not be successful if it was limited to the local plants and concerned only with the unionized workers and their management.

The second phase in the process concentrated on bringing the middle management group of the operation into the process. With some successes to point to at the plant level, we found it possible to move somewhat faster in the second phase. The techniques were not all that different, however. A series of business conferences and workshops was organized for teams that made up the total business group. These meetings involved middle managers who shared common issues, for example, management teams from plants that produced similar products and the entire executive body from division headquarters who oversaw that constellation of plants. This provided an opportunity— even the necessity—for a role reversal with the plant managers now modeling and sharing with the headquarters "bosses" what they had learned at the plant level. The headquarters personnel, in turn, found themselves responsible for articulating the business vision and competitive outlook for the enterprise as a whole, thus binding the plant management teams to the goals of the larger enterprise.

The meetings were held off-site in an environment of open discussion, free exchange, and constant debate—always with a coordinator who had an earlier successful experience in the process. Typically, five days of continuous immersion in the business problems was required. The results of these meetings might be briefly summarized by one phrase—improved understanding. The participants came away with an improved understanding of the business issues, their customers, the competition, the problems and constraints faced by others in the organization, and the potential benefits of the employee involvement process. Further, the participants discovered that the opportunity to discuss and debate the issues in this setting opened a wealth of new communication channels.

More lessons were learned from these experiences. Two were perhaps most important:

1. Involvement of the entire "system" in the process greatly reduces the time needed to accomplish a change in culture.
2. Debate is necessary to accomplish change—it both clarifies the issues and is an essential element in obtaining genuine commitment from everyone as to future actions. The building of this genuine commitment and understanding stood in marked contrast to the shoulder-shrugging compliance that had, in the past, been mistaken for agreeing with "the boss."

Phase III—Senior Management Change (1985–present)

As the process proceeded and included more and more people throughout the organization, it became clear that a continuing mechanism was needed to provide a meaningful involvement of the senior management in this process. This led to the establishment of the Ford Executive Development Center. The center has the responsibility to provide an environment for bringing together the company's executives from around the world to develop a consensus on company strategies and to focus on the customers and their needs and wants. The format of the group sessions is similar to that used in the earlier stages of the program—a site away from the office, groups of 50 or less, a mixture of people from various levels of the organization worldwide, participation throughout the five days by at least three of the corporate vice-presidents, and a concluding session in which company issues are debated and discussed with the chairman or the president of the company. During the week, speakers from outside the company are also invited to challenge the group to think differently about particular problems.

As we found in the earlier phases of this effort, the lessons we learned are not revolutionary. It is critically important, however, that we continue to operate in such a way that those lessons are not forgotten. Perhaps the three clearest lessons from Phase III are as follows:

1. The executive development center provides a straightforward means by which senior managers can acquire a worldwide outlook about the business.
2. A greater openness has developed among Ford employees to new, even disturbing concepts necessary to our success in the future.
3. Debate and questioning of the status quo with the senior management is an important part of understanding and developing support for the worthiness and rightness of corporate strategic objectives.

CONCLUSIONS

An effective work force that is encouraged to search for ways of achieving continuous improvement is a key foundation of the modern manufacturing enterprise. Employee involvement in problem identification and employee empowerment that encourages the employee to take actions in creating a more efficient and effective system are critical to becoming world class. While it is not easy for large organizations to be "people oriented," their ultimate success in meeting the demands of the marketplace requires a commitment and a willingness to search for new ways of solving problems. "Employee involvement" is one of the keys to achieving this. The process is never ending. Employee involvement must be constantly stressed, continuously practiced, and regularly evaluated if it is to become a foundation of the company's operating philosophy.

Implementation Projects: Decisions and Expenditures

H. KENT BOWEN

Measures of the quality, cost, timeliness, dependability, flexibility, and innovation of a manufacturing system are based on customers' views of the product and service. The customer's view is determined by what happens in the other functions such as design, production, and vendor procurement. Absent equations, models, and simulation systems, we look to empirical observations—data, trends, and recurring events—to stimulate fundamental learning. The challenge in determining the manufacturing system outcomes (e.g., enterprise growth, market share growth, profits, or lifetime employment) and their relationship to the system measures will require substantial effort. I have chosen one type of empirical observation in this chapter to indicate the impact that foundations may have, once understood. The case we will consider here is the new product or process introduction project—decisions and their time-related costs.

A knowledge-based work environment and rapid product and process introduction cycles require a close linkage between the planning and designing and the implementation and execution. As products and processes become inherently more complex and technology driven, the absence of clear objectives and of ready-to-access skills, knowledge, and technology results in a search for cause-and-effect relationships. What is the relationship of knowledge gained from previous products and processes to that gained from new products and processes? How do the early decisions (such as product design features, process selection, and work force selection) link to future events once the development and implementation process begins.

In efforts to study new product or process introduction projects, I have made a set of empirical observations that provide understanding of one aspect of the manufacturing system. These observations may allow mental models of how these processes work and thus suggest opportunities for improvements to a key feature of time-based competition—rapid and effective introduction of new products or processes.

The empirical observations seem to be valid for many U.S. manufacturers, irrespective of the particular market or technology. Similar data have been reported for the introduction of products or processes related to commercial aircraft, computer workstations, high-density information storage systems, cellular phones, and advanced materials and components.

The details of each example are less important here than are the trends. The starting point, $t = 0$, of a commercialization project occurs after initial research and development have determined a significant value to, and acceptable risk for, moving the development into an implementation or commercialization phase. The technical feasibility of the product or process has been verified, but there are still uncertainties that require refinements, integration functions, and pilot developments. The project gets a title, resources are allocated, and someone is assigned to track the expenditures; most often there is a carefully constructed schedule with gates, hurdles, or phase reviews.

For many of these commercialization projects, the schematic representation of the accumulated expenditures shown in Figure 1 is universal. A slow ramping of expended dollars or engineering hours occurs during the initial 40–70 percent of the elapsed time. The early period is mostly expenditures of people on design and primitive prototypes. The later rapid rise is associated with purchases and installation of equipment and facilities and the training of the operators. The rates are quite different in these two periods. The teams involved in these projects perceive the actual time of the decisions that triggered the expenditures as occurring very early in the process. This perception is shown schematically in Figure 1, and I have found that within 15–20 percent of the total elapsed time, 80–85 percent of the key decisions about future expenditure have occurred. Complex processes involving numerous variables and elements of subsystems (such as information, technology, human, financial, and marketing) result in longer than anticipated execution consequences and, thus, strongly influence feedback loops in the manufacturing system. We will call this observed relationship between essentially sunk costs and decisions the 15/85 rule.

The 15/85 rule poses many questions about how projects are managed:

1. When is the appropriate time for senior management to become involved in new product and process projects?
2. What should the leadership of such teams be, and are there opportunities to change leaders midstream?

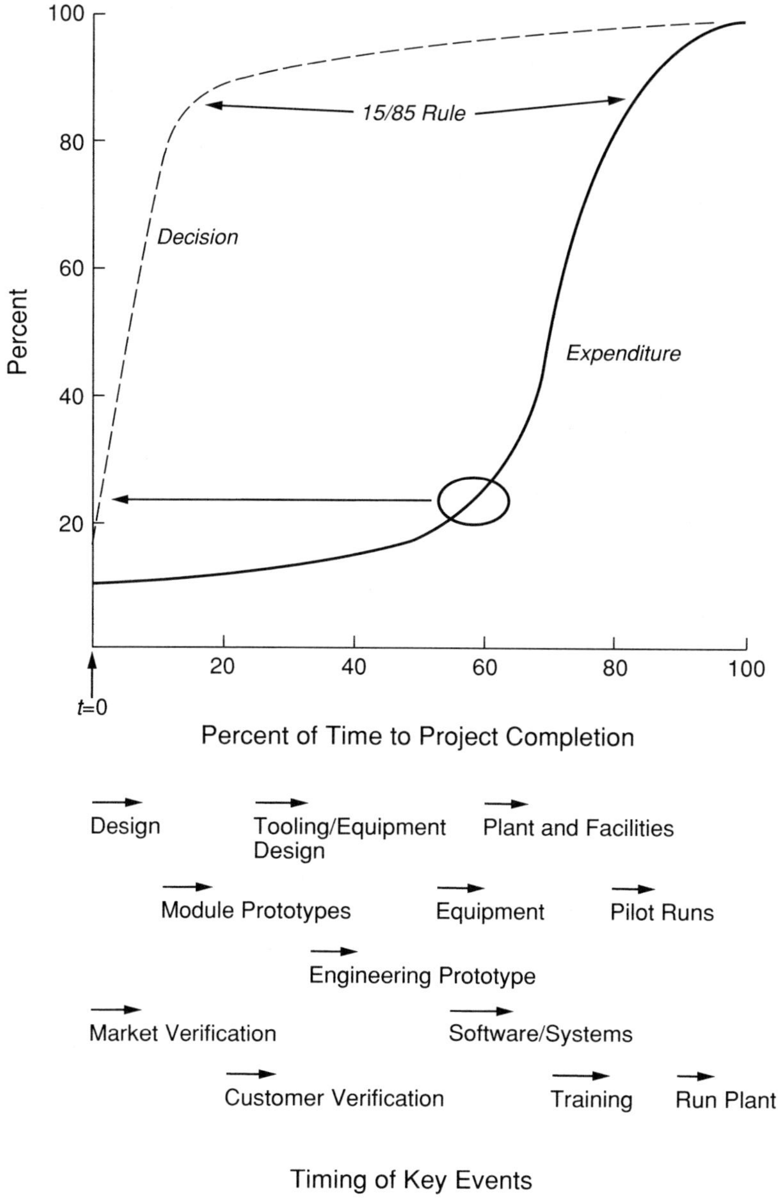

FIGURE 1 Schematic relationship between project expenditures and the decisions about the product or process.

3. How much of the loss due to engineering change orders, redesigns, and the like is due to the long cycle times (decision to implementation), and how much is a result of uncertainty about the component part and system that is being designed, built, or implemented?
4. What is the value of knowledge (know-how and know-why) at the early stages, that is, before 15 percent of the time has elapsed?
5. How should project teams be staffed, organized, and managed to promote organizational learning as well as to accomplish project objectives?

There have been numerous recent examples of a large disparity between companies doing similar projects. Plain paper copiers, workstations, or automobiles are product examples for which careful studies have shown as large as twofold differences in the elapsed time and the engineering work-hours or expended funds when implementing similar projects. There are suggestions from the cases of the best-of-the-best that the 15/85 rule does not apply; the decisions are much more closely linked to the doing and the expenditures. The feedback and corrections are different in number, timing, and quality. The most comprehensive comparisons of product development projects is that of Clark and Fujimoto (1991), who also link these processes to key manufacturing metrics such as the quality and rate of innovation of the product.

What are the common sources of delays and causes of rework and design changes that result in the 15/85 curves? There are organizational aspects, such as the ineffective working of teams pulled from functional groups. There are systems considerations, such as the lack of standards or a single data set. Other aspects of the problem include procedures and mechanisms for problem solving and structuring of the solutions. In addition, there are infrastructural issues such as lengthy procedures and justification for obtaining resources—people or capital.

All of these sources of delay have been observed to a greater or lesser degree in new product and process introductions. Further analysis of Figure 1 allows us to emphasize another common element in 15/85 style projects. The issue is how uncertainty diminishes with time as the project proceeds.

In an absolute sense the standard for elapsed time in a project is often derived from ad hoc or artificial means, a $t = 0$ point is assigned arbitrarily when a schedule is articulated. Based on experience and the tasks to be achieved, the schedule anticipates preparation and completion of certain events. These events include milestones and reviews, but most often a scheduled event may trigger other events that cause activation of resources and consequentially expenditures. In many cases the schedule dictates these events rather than the events being triggered by accomplishment.

Figure 2 continues the illustration of the expenditure of effort and deci-

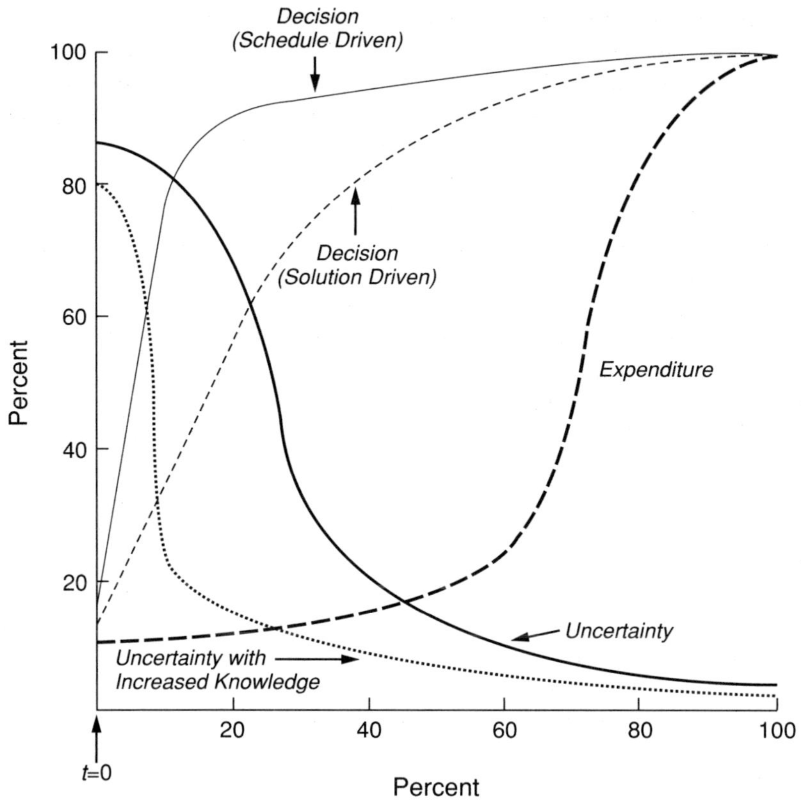

FIGURE 2 Representation of the effects of early knowledge and problem solving (reducing uncertainty) on the changed relationship between decisions and expenditures in projects.

sions but with the added dimension of uncertainty. Here uncertainty includes anything that may cause an element or the integrated whole not to meet the required specifications. The new product may require unproven process technology; and thus without small-scale and production-scale tests of the process, the success of the project is still uncertain. Guesses about facility configurations and yields, for example, become the decisions that will generally require engineering fixes at a later date. Later fixes usually require more time and larger expenditures than corrective actions taken early before other processes, designs, and systems became determined and influenced by imperfect data.

The uncertainty about the new product or process is of course subject to customer and market uncertainties; but let us focus on uncertainties in the technology (hardware, software, and systems for designing, testing, or making) and in the people, organizations, and structures (knowledge, skills, methods, and systems) directly applicable to the project. In short, individual as well as organizational knowledge and the skill with which problems and opportunities are identified and solved dictate the rate at which uncertainty is reduced. Should "inventions" be required after $t = 0$? Since real inventions, as opposed to innovations, cannot be scheduled, commercialization projects in which time is a determinant require that the inventing be done before the start of the project. Thus, the "planned" uncertainty has to do with applying the technology, integrating the elements, and extending or incorporating known aspects from other realms into this particular project.

The schematic diagram in Figure 2 shows what occurs in 15/85 projects when the project is driven by problems solved (uncertainty reduced) rather than dogmatic adherence to a predetermined schedule. A more rapid decrease in the uncertainty of a whole and its component parts for a new product or process allows the decisions to be more closely linked in time to the expenditures—the doing. Projects focused on problem identification and solution to increase knowledge and reduce uncertainty use processes such as project reviews, prototyping, pre-project understanding, cross-functional teams, and benchmarking as drivers for decision making and resource allocation. For example, component prototypes are done early and rapidly. Production-scale engineering prototypes are tested before half the time has elapsed in order to allow for refinements and improvements at the system level.

What is the value of the "right knowledge" early in the commercialization project? Extensive research will be required to quantify the answer to this question, but the nature of the empirical observations represented in the figures suggests multiples of 10, and perhaps as much as 100 to 1,000, in savings. The cost and consequences of early, ill-informed decisions put into place processes (expenditures) that can be measured and judged only much later.

As important as the need for knowledge is the type of knowledge needed. For example, the pre-project knowledge for a new product must come not only from the advanced development lab but also from the factory floor and process development lab. Knowledge about the capability of individual and linked processes and process steps, of machines and process lines, and of integrated production systems is a key to reducing uncertainty because it establishes current and future possibilities for the new product. This understanding sets the agenda for problem resolution when more advanced processes are required.

The special requirements of added process and factory knowledge is exacerbated in time-based competition. The availability of design tools and methods has improved the product design aspects of manufacturing faster than our knowledge about developing and installing new production technologies. The pacing element is becoming the new product production system and not the new product design—time-based competition is process knowledge competition.

The manufacturing enterprise, the manufacturer, is a system. One central activity, introduction of a new product or process, is itself a system with all the risks and uncertainties of complex systems. When considering the enterprise or any element of it, understanding becomes richer when one looks for the interconnections of activities, functions, processes, and outcomes.

In this chapter we have looked at empirical observations that relate the expenditure of resources for introducing a new product or process to the timing of decisions that eventually caused the expenditures. The system aspects of developing and commercializing new products and processes require linkages and feedback loops to discover and solve problems. This is more than concurrent or simultaneous engineering (Nevins and Whitney, 1989) or quality engineering (Phadke, 1989). The consequences of the empirical observations are many. Experiments to validate cause and effect are suggested, mental models can be conceived and tested through field studies, and quantitative relationships could be estimated. Clark and Fujimoto have shown that the methods and practices of project teams (formal and informal) and the way in which teams are organized affect the performance measures. I see fruitful research, much speculation, and the emergence of a foundation for the manufacturing system.

Benchmarking

W. DALE COMPTON

Achieving and maintaining a competitive position in the worldwide marketplace demands an awareness of the capabilities of one's competitors and the ability to continuously improve one's relationship to the leaders. To measure oneself against the world leaders of the competitive race is an essential element of good management. As Camp (1989) notes, the Japanese describe this by the word *dantotsu*—"striving to be the 'best-of-the-best.'" Camp observes that "We in America have no such word, perhaps because we always assumed we were the best." The quantitative comparison of one's current performance against the world leader is the essence of "benchmarking."

The metrics that are useful for benchmarking are numerous and varied. In some industries it is common to emphasize a few operating characteristics, while in others the focus will be on a wide range of parameters. In some cases the parameters that describe system performance will be highlighted. In others, it may be more common to examine the performance of subelements of the system. Some metrics may be reasonably easy to obtain. Others may need to be estimated based on limited available data. However it is accomplished, benchmarking is a critical foundation for the successful operation of a manufacturing enterprise.

We will enumerate several of the metrics that are commonly used in making financial, operational, and system-oriented comparisons. Each measures a characteristic that is important to an operation. When properly aggregated, they offer a means of assessing one's overall competitive strengths and weaknesses.

FINANCIAL METRICS

A wide variety of financial metrics are commonly used in assessing the competitive performance of an enterprise. Although many of these can be obtained, or estimated, from the annual financial reports of the enterprise, the lack of disaggregation of data on the part of many firms often prevents a detailed analysis of the individual business operations. In the United States, the Securities and Exchange Commission requires certain reports that are much more detailed than are found in most annual corporate reports. The "10K" report, for example, contains a variety of information that can be used to derive comparisons with domestic competitors. Even though reports by foreign competitors are often less detailed than are those of U.S. firms, privately held firms often publish no details. Assessing the performance of such companies is often difficult.

The following metrics are useful in comparing the financial performance of enterprises:

- Return on assets
- Return on sales
- Return on investment
- Unit cost of product
- Fraction of unit cost of product resulting from:
  - Labor
  - Materials
  - Capital investment, etc.
- Profit per unit product (Averaged over all products offered by the firm)
- Profit per unit for each product type
- Number of employees
- Average sales per employee
- Average hourly wage of each employee
- Average hourly benefit cost for each employee
- Cash flow
- Equity ratio
- Annual volume at which the break-even point occurs

PRODUCT PERFORMANCE METRICS

The performance of the product in the hands of the consumer is a key indicator of the capability of the development, production, and marketing system. The successful manufacturing enterprise does not, however, simply wait to see emerging trends in market share and then respond. A constant assessment of the relative merits of competing products and the continuous incorporation of the appropriate responses into one's products are a mark of the successful firm. The "appropriate response" must be determined, of course, within the context of the *needs and wants of the customer*. The following are some of the *actions that successful enterprises have institutionalized* to obtain this information:

- Evaluation of competing products under conditions similar to those used to test your products
- Tear down analysis of competing products
- Reverse engineering of competing products
- Customer surveys of competing products

From tests and evaluations of the type described above, one obtains information on the following important metrics:

- Part counts
- Material types used
- Material utilization in each component
- Processes used in production, e.g., assembly techniques
- Product costs
- Service capability, e.g., field repair versus field replacement
- FMEA (failure modes effects analysis)

From customer surveys undertaken either directly or through industry-wide surveys carried out on behalf of the industry, one can assess:

- Quality of the product as experienced by the customer
- Long-term durability of the product
- Fraction of sales to repeat customers
- Responsiveness of the producer to service requests

UNIT OPERATION METRICS

A manufacturing enterprise must be concerned with performance at many levels: the unit operations employed in producing the product; the consolidated operations as a manufacturing system; and the relationship between the manufacturing enterprise and the other elements of the firm, such as engineering and marketing. Whether achieving continuous improvement in an existing operation or planning for upgrading or replacing a facility, the choice of the unit operations, their arrangements, their interactions, and their controls is critical to determining the overall competitive position of a manufacturing operation. A variety of metrics can be useful in making these assessments:

- Time required to accomplish a unit process
- Time a product spends in a process versus nonproductive time spent in waiting and setup
- Buffer sizes used for each unit process
- Machine reliability
- Yield and quality of unit processes
- Machine utilization rate
- Amount of material scrap
- Labor hours per unit process
- Energy use per unit process
- Time required to change a process, e.g., a die change

Determining these metrics for a competitor may be very difficult. Companies have used a variety of methods to obtain these data. Vendors who supply facilities to both you and your competitors can often provide insight into the efficiency with which others use particular machinery or processes. Trade associations can be enlisted to provide surveys in which all members of an industry can participate (*Textile World*, 1989). Foreign partners and subsidiaries often can provide information on the state of operations in other countries. Professional society meetings and conferences provide valuable opportunities to exchange timely information in a neutral environment, to the benefit of all parties.

SYSTEM OPERATIONAL METRICS

While one must be concerned with the metrics of the product and with those that describe the unit operations, a clear vision of the operation of the total enterprise cannot be obtained until the analysis is broadened to assess the operation of the *total* system. It is in this context that one can begin to understand overall performance relative to world-class producers. The following are typical of the metrics needed to determine system operation:

- System productivity
- Units produced per hour of labor
- Units produced per total investment
- Hours of direct labor per unit versus total system hours (management, staffs, etc.) of labor per unit
- Quality of the operations
- Rejects during processing
- Field repairs per units delivered
- Returns by customers
- Inventory turns
- Total work-in-process
- Value of work-in-process

- Fraction of time facilities are used
- Fraction of production facilities that are new or fully depreciated
- Extent of uniform use of all unit operations
- Time required to respond to a changing market demand
- Time required to introduce a new product or service
- Extent to which just-in-time methods are employed
- Extent to which concurrent/simultaneous engineering is practiced

AGGREGATED MEASURES OF PERFORMANCE

In addition to the above metrics, the successful competitor must be concerned with some aggregate characteristics of the enterprise. Many of these are useful in characterizing the dynamic response of the system to changes in the environment. The following list of some of these characteristics includes a brief indication of a few of the common definitions of these measures. Obviously, various metrics will be needed to measure the performance for the alternative definitions of the following characteristics:

- Flexibility—(i) the capability of the system to respond to a changing marketplace, (ii) to accommodate new technology, or (iii) to reflect changing workplace practices.
- Complexity—an aggregate property resulting from the separate decisions of the functional departments of the enterprise.
- Variability—can be interpreted as (i) random variations in processes, (ii) the temporal variations of a demand, or (iii) the "variety" in the products or processes.
- Reliability—used in the sense that a product, process, or system that is subject to failure can be expected to perform as required.
- Quality—(i) a transcendent property, (ii) a product-based property, (iii) a user-based property, (iv) a manufacturing-based property, or (v) a value-based property (Garvin, 1984).
- Availability—(i) the fraction of time that an item is in inventory and is available for shipment on demand or (ii) the mean fraction of time a machine is available for operation.

MEASURES OF LONG-TERM COMPETITIVE CAPABILITY

The above metrics are generally viewed as describing the current status of an enterprise. As important as these are for comparing a firm's current capabilities with those of a competitor, they do not offer a clear view of the future capabilities of the firm. Two areas are particularly important for a firm that is competing in the world marketplace—its technological capability and the quality of its personnel.

Technological Capability

Recognizing that the variety and breadth of technology that most enterprises employ is so great that they cannot hope to develop it entirely within their own organization, it is essential that they maintain mechanisms for identifying promising technological developments and for adapting them to their particular needs. Means of accomplishing this include the following actions:

- Support of R&D projects within the organization
- Development of pilot lines for investigating new processes
- Joint investigations of technological opportunities with vendors
- Participation in industry consortia that are developing new technologies
- Involvement with selected university research activities
- Regular attendance by personnel at professional and technical meetings

The metrics by which an organization should judge itself against its competitors include the following:

- Level of support for internal R&D. Fraction of R&D budgets devoted to long-term projects; level of investment in exploratory opportunities, including pilot lines, new product concepts, and technological innovations.
- Support provided personnel to participate in worldwide technical meetings and forums.
- Level of support for technical libraries and information systems.

Technical Personnel

The capability of an organization to monitor and identify technological opportunities depends critically on the presence of well-trained, technically astute personnel. Some of the metrics that are useful in assessing one's capability relative to competition are

- Fraction of technical work force with professional degrees
- Fraction of technical work force with advanced degrees
- Fraction of technical work force regularly involved in continuing education
- Level of recognition that has been afforded technical personnel by outside organizations, e.g., honors, membership on national committees, officers of professional societies, invited lectures
- Level of involvement of technical personnel in assessment of opportunities for new products, processes, or services

INTERACTIONS

It must be recognized that many of the metrics identified above are not independent. For example, it is clear that the performance measured by financial metrics depends directly on many of the product, process, and system metrics. Similarly, the actions that determine the quality metric (depending on which property one is measuring, of course) have an impact on a host of other metrics, for example, process yield, product costs, use of facilities, levels of employment, material scrappage, labor hours per process, and fraction of repeat sales to customers. It is evident that care must be taken in interpreting a combined list of metrics.

Perhaps of more concern are the indirect relationships that may exist but are not easily quantified. For example, how does one express the relationship between the quality of the operations or the level of system productivity on flexibility and complexity of the operation? The absence of a clear understanding of these relationships—beyond some general rules of thumb—presents a serious barrier to understanding the likely impact of actions that may affect more than one part of an enterprise.

CONCLUSION

The conclusion that is to be drawn from the above discussion is that there is a wide variety of metrics—many more than those listed here—that need to be regularly and consistently measured before firms can assure themselves that they are performing competitively. Benchmarking is an important element of the foundations of manufacturing. Without good benchmarks a firm cannot be assured that its current level of performance is appropriate. Neither can it be assured that the objectives established for future improvement will be adequate. Good data are essential to good decisions. A well-established practice of benchmarking offers the best opportunity for generating those data.

Improving Quality Through the Concept of Learning Curves

W. DALE COMPTON, MICHELLE D. DUNLAP,
and JOSEPH A. HEIM

Quality is the hallmark of competitive products. Consumers reject products that are of inferior quality, and they shun companies who are perceived to provide products or services with less than competitive quality. A company cannot survive in the current world marketplace without providing a product or service that is of high quality. It would be hard to find a U.S.-based manufacturing enterprise that does not place quality near the top of a list of strategic or operating objectives that would also include cost, innovation, and customer focus.

This sensitivity to the customer demand for quality has not always been a dominant force in the operating strategies of U.S. companies. Having realized its importance, companies find that they must now direct their energies in ways that focus on this objective. A broad range of operating procedures must be modified or, in some cases, created: quality must be designed into products; a productive interaction must be stimulated among the design, manufacturing, and marketing activities; input from the customers must be obtained and used; and active participation by their employees in creating "quality" products and services must be encouraged.

With this new awareness of the importance of quality have come two specific needs. First, there is a need to measure the performance of the organization against that of its competitors and, second, there is the need to assess the trends in one's performance in order to take appropriate actions to ensure continuous improvement. In the first case, an absolute measure of performance is needed. This is sometimes referred to as "benchmarking," or measuring oneself against the world leader—the "best-of-the-best"—in a product or process arena (Compton, in this volume). In the second, progress over time is the prime concern, that is, how well the organization is achieving continuous improvement in performance. A proper combination of these two measures is critical. Without them an organization cannot properly evaluate its absolute competitive status, nor can it be assured of its ability to remain competitive over a long period of time. With "high quality" as a prerequisite for being competitive in the marketplace, achieving and maintaining high quality in all aspects of an operation is a critical component of the foundations of all manufacturing systems.

ORIENTATION

This article focuses on the second of the two needs identified above, assessing how an organization improves over time. We will be concerned, therefore, with assessing trends in quality. We will conclude with some observations about the need for continuous and careful collection of the type of data that are critical to a proper assessment of progress.

"Quality" is not a universal descriptor that has a unique definition under all circumstances. Garvin (1984) has described five approaches to defining quality. The appropriate metric for measuring the quality of a product or process will depend on the definition or circumstance that is of immediate interest. It can, for example, refer to defects arising from a production process, defective parts shipped to customers, or reliability of the product in service. Although we will not discuss the various measures of quality in this paper, we have obtained examples of each of the above measures. We offer examples of the first two in this paper.

Measures of the quality of the outputs of a system can be obtained in many ways. In the day-to-day operation of a manufacturing enterprise, the collection and use of process data to support statistical process control (SPC) is important to achieving high-quality manufacturing. SPC requires that measures of one or more attributes of the quality of a production system be regularly employed, and it provides a paradigm for the efficient use of those measures to control the process. There is ample evidence of the importance of this real-time control in improving the quality of the processes and the products that result from these processes. In every sense, the effective use of SPC and total quality control (TQC) have become important elements of the foundations of effective manufacturing systems.

The measures of quality that are implicit in the application of SPC necessarily concern shorter time periods; that is, they reflect the current status of the process or system that is producing the product. Although the importance of this near-term collection of data—and appropriate analysis to accomplish SPC—is not questioned, it is also clear that an understanding and quantification of the longer-term trends in quality are also critical to achieving continuous improvement in quality.

THE LEARNING CURVE RELATED TO COSTS

A traditional approach to measuring the long-term cost performance in a manufacturing operation is to use the "experience" or "learning" curve concept (Henderson and Levy, 1965). This asserts that the fractional reduction in the average cumulative cost (in constant units of measure) of producing a product is proportional to the fractional increase in the quantity of the product that is produced and yields a power law representation that is similar to that first described by Wright (1936). A common formulation of this law relates the cost of production of the nth unit, X_n to the total production volume N

$$X_n = KN^{-b} , \tag{1}$$

for large N.

Equation 1 has been used many times and has been shown to be valid for a wide variety of products in many different industries (see Argote and Epple, 1990, for a discussion of this form of the learning curve in manufacturing). The literature contains numerous discussions of circumstances in which an exponential law is the appropriate formulation for the learning curve (Buck et al., 1976; Pegals, 1969). A simpler formulation for a learning curve, seldom used in the literature and applicable only under limited circumstances, is the linear representation. Determination of the form that is most appropriate depends on many factors, including the nature of the data sampling protocol. In general, however, if it is not possible to determine which form is most appropriate, either because of an absence of a priori knowledge or because of a lack of sufficient high-quality data, the simplest formulation is probably best. Selecting the simplest formulation entails testing to determine whether the data are best fitted by a linear, an exponential, or a power law representation.

Irrespective of the formulation chosen, learning curves are not to be viewed as merely descriptive. They can be, and frequently have been, used as an aid in making predictions, in that early experience in the production of a product can be used to predict future manufacturing costs. Assuming that one has confidence in the form of the equation that is chosen—whether power, exponential, or linear—and that one can make a reasonable estimate of the constants that appear in them, one can readily predict the costs to produce a unit after some future cumulative production volume has been achieved. Even in the absence of detailed data on a given product, the experience of many manufacturers with many products is that manufacturing costs can be expected to decrease by 10 to 20 percent for each doubling of production volume. Abernathy and Wayne (1974) have explored the limits of validity of the learning curve concept.

The improvement depicted by the experience curve is a result of conscious effort and attention on the part of the management and employees of the enterprise. It cannot be expected to continue without the attention and focus that accompanies a clearly accepted operating objective, in this case an objective of continuously reducing the costs to manufacture the product or to offer the service. A variety of actions combine to produce the desired cost reductions (Allan, 1975):

1. Improved efficiency in the use of labor through training and incentives.
2. Introduction of new and improved processes that reduce manufacturing costs.
3. Redesign of the product to reduce manufacturing costs.
4. Standardization of the product to reduce the variety of tasks demanded of the workers.
5. Scale effects resulting from large volume production.
6. Substitution of lower-cost materials while retaining product features.

THE LEARNING CURVE RELATED TO QUALITY

Just as competitive pressures have forced the management of U.S. companies to pay special attention to costs, so also is management being forced to pay special attention to improvement in quality. Although many approaches are taken to improve quality, these efforts have a few key actions in common:

1. Simplification of product design to enhance manufacturability.
2. Involvement of the employees in designing the manufacturing system.
3. Enhanced training of the employees.
4. Substitution of automated machinery in areas that are not conducive to human operation.
5. Collection of extensive data on each operation, and analysis to identify problems and trends in those operations.
6. Introduction of new or improved processes that are less sensitive to variation.

Although the specific actions taken to improve quality differ from those taken to reduce unit costs, a striking similarity exists between the two lists. In particular, both result from conscious actions taken by management and employees to accomplish a common strategic objective for the enterprise.

Both combine human commitment and training with technical improvements. Both require extensive knowledge of the processes being employed and the products being produced. Therefore, quality and costs might be expected to share a common representation. One might then speculate that quality should follow an experience curve similar to that of cost. By analogy, therefore, a quality learning curve might take one of three forms such that the quality index (QI) for the *n*th item is defined as follows:

Power law: $$\mathrm{QI}_n = (\mathrm{QI})_1\, N^{\pm m} \quad (2)$$

Exponential form: $$\mathrm{QI}_n = (\mathrm{QI})_n^{*} + (\mathrm{QI})_o\, e \quad (3)$$

Linear representation: $$\mathrm{QI}_n = (\mathrm{QI})_a \pm (\mathrm{QI})_b\, N \quad (4)$$

In the above equations, $(\mathrm{QI})_n^{*}$ is the asymptotic value of the quality index, $(\mathrm{QI})_o$ is related to $(\mathrm{QI})_1$, the quality of the first unit produced and to $(\mathrm{QI})_n^{*}$, $(\mathrm{QI})_a$ and $(\mathrm{QI})_b$ are constants, and N is the cumulative volume of the products produced. In Equations 2, 3, and 4, the sign can be positive or negative—positive if the quality index is improving as cumulative production volume increases, for example, yield from a process; negative if the quality index reflects defects or defective parts, which will decrease as the cumulative production volume increases.

While the particular attribute of the product or process being considered will most likely be different for each product and process, the above formulations are independent of the specific attribute that can be related to the quality index. One should not expect, however, the numerical values of the constants to lie within a specific range or to have any particular relationship from one product to another, because the quality indices can differ depending on the attribute chosen for examination.

OBSERVATIONS

Schneiderman (1988) appears to be one of the first to treat production yields or the quality of products shipped according to a learning curve. Schneiderman offers a number of examples of quality learning curves that are presented as exponential formulations in which a measure of quality is plotted as a function of time from the start of production. It should be noted that this formulation is consistent with Equation 3 only in the case that production rates are constant over time—a circumstance that seldom occurs.

A test of the hypothesis that a quality index is describable by Equations 2, 3, or 4 can, in principle, be made by examining the quality of products or processes at various levels of production. For some dozen products—for which measures of quality and production volumes could be obtained—we have generally found that two of the three formulations are virtually indistinguishable in their ability to represent the data. In some cases, the linear and exponential laws were indistinguishable—meaning that the coefficient of correlation for the two laws was nearly the same—while in others, the power and exponential laws were indistinguishable. We found no case in which all three representations were equally good.

Data are presented in Figures 1 through 3 relating an index of quality to the cumulative volumes of production for three different products—light bulbs, a small electric motor, and grey iron castings. General Electric Company and The Dalton Foundries, Inc., graciously supplied the data contained in these figures.

A description of the quality index for each of the products is given in the figure captions. Having no a priori basis on which to choose the preferred formulation for representing the quality index, we examined each of the products using Equations 2, 3, and 4. Following an observation by Buck et al. (1976) that the exponential form of the learning curve is some-

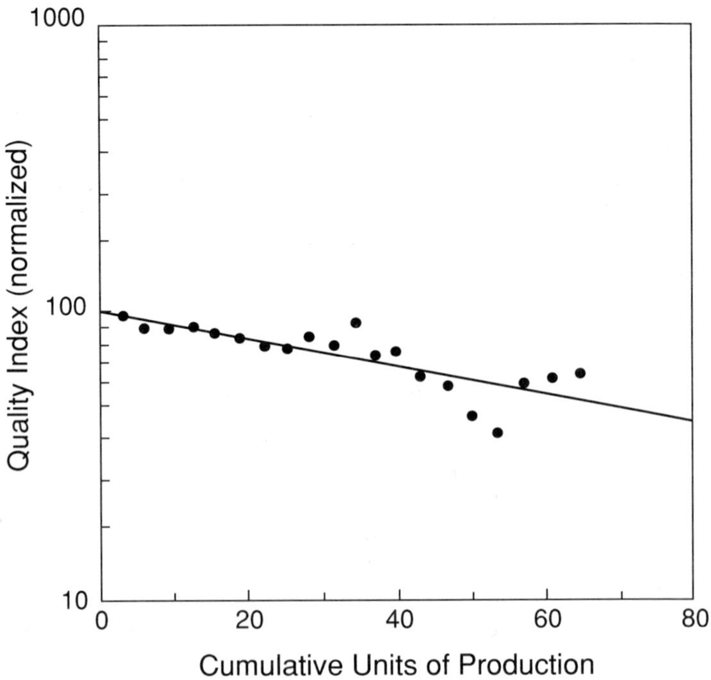

FIGURE 1 A normalized measure of defects in light bulbs for the period between 1970 and 1989 as a function of normalized volumes. Quality is measured in defects per million light bulbs. These data represent the cumulative production from a single plant. Correlation coefficient r = .82 (for the linear representation, r = .86). Courtesy of General Electric Co.

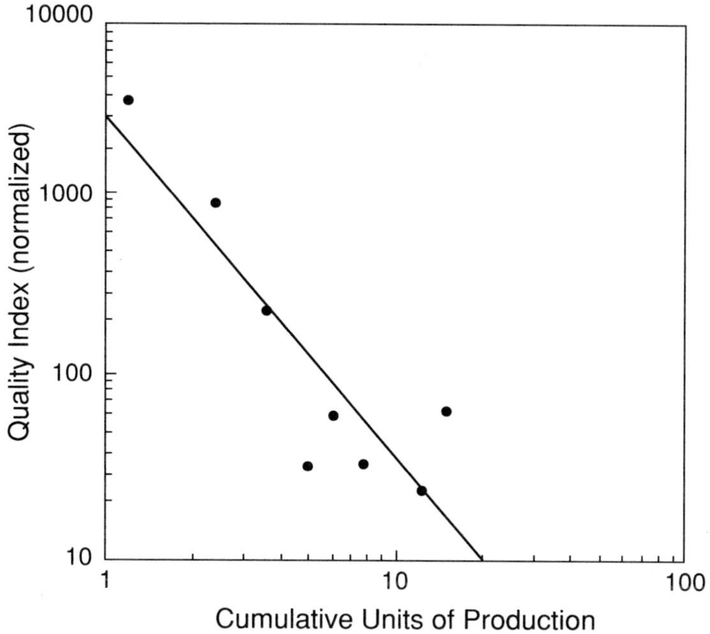

FIGURE 2 Normalized defects as a function of cumulative volumes, in arbitrary units, for small electric motors for the period 1981 through 1989. Defects due to production errors are in the range of a few tens per millions of motors. Correlation coefficient r = .89 (for the exponential representation, r = .70). Courtesy of General Electric Co.

what to be preferred for batch, or average, sampling of the metric in question, we have chosen to present three of these sets of data in terms of the exponential relationship of the quality index to the cumulative volumes of the product produced. In none of these cases was saturation apparent, implying either that $(QI)_n^*$ was effectively zero or that the observed values are so far from the saturation value that the present representation is not adequate to display a saturation. It is of particular interest that the correlation coefficients for a linear plot of the data shown in Figures 1 and 3 are essentially the same as shown for the curves as plotted. The correlation coefficients for the curves shown in the graphs are given in the captions, along with the correlation for the best alternative formulation. Each of the data points in these three figures represents an average of the quality metric for a period of one year. Thus, for Figure 1, the quality data are for 20 years of production, Figure 2 for 9 years, and Figure 3 for 12 years. In some cases the index is defined as defects in production; in others, the shipping of a faulty product to a customer.

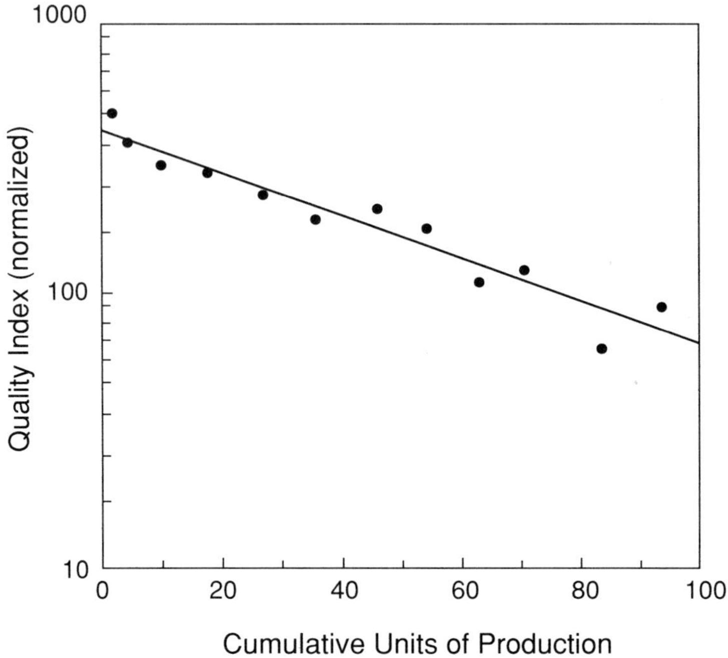

FIGURE 3 Normalized scrap rate as a function of cumulative volumes for the period 1979 through 1989 for a grey iron casting. Correlation coefficient r = .96 (for the linear representation r = .94). Courtesy of The Dalton Foundries, Inc.

CONCLUSION

In each of the cases examined, there is clear evidence that the quality index, although defined differently for each group of products, is related to cumulative volume of production. With the diversity in product type and processes represented here, the hypothesis that one or more forms of the learning curve exists for quality is supported by these data. Although the present data do not appear capable of distinguishing among the various forms for a learning curve for quality, it appears that one or more forms can easily be found to permit a reasonable extension for setting new goals or examining the impact of past actions on performance.

Because the time frame in which these products were in production is long, it is reasonably certain that the actions suggested earlier as being important for management and employees in achieving a continuous improvement in quality were taken throughout the life of these products. Each of the companies that provided these data has indicated that the quality trends demonstrated in these figures are the result of constant and consistent attention to the importance of continuous improvement. Although this dependence of a quality index on cumulative production has been demonstrated for only this select group of products, we believe that this phenomenon is generally true. A collection of additional examples from other industries would help support this conclusion.

We have been surprised to find that few companies keep data in the form or with a consistency that allows the following of trends as described in the types of curves shown here. In our view, this is a shortcoming that should be addressed by all concerned with continuous improvement. The systematic collection of data on quality and the representation of these data in the form described by Equations 2, 3, or 4, offer a means of tracking progress on the "continuous improvement" of quality and a means by which realistic expectations can be established for future goals. Above all, the existence of a learning curve for quality should be viewed as one more example of the need for careful collection of systematic data. Without good data, this important foundation cannot be used.

Creating high-quality products through high-quality manufacturing processes and systems is a critical element in the foundations of manufacturing. The learning curve for quality should therefore be viewed as an important element in the foundations of manufacturing. We believe that the learning curves can be an important contributor to achieving improved quality.

ACKNOWLEDGMENT

This work was supported in part by a grant from the Ford Motor Company Fund.

Organizing Manufacturing Enterprises for Customer Satisfaction

HARRY E. COOK

There is considerable dissatisfaction today with the so-called functional organization, as shown schematically in Figure 1. It is not so functional anymore because it tends to support a sequential manner of product realization, which is believed to be a significant source of substandard cost, quality, and lead-time performance versus enterprises that operate in a more parallel fashion (Clark and Fujimoto, 1989b; Dertouzos et al., 1989; Hayes et al., 1988; and Stalk and Hout, 1990). However, before the onset of highly competitive, global markets, the functional organization seemed adequate to the task. In markets that were only weakly competitive, the enterprise could move slowly and still be successful. It could control quality by inspection. It could control costs by having design engineers measure their results against a design cost standard based on a process unlike the one that would actually be used to make the part.

In searching for an organizational structure that better suits today's highly competitive environment, it is useful to have a means of forecasting the effectiveness of a structure under consideration. The obvious approach is to look at the most successful companies, see how they are organized, and adopt their structure. However, organizational structure is not the only factor that determines the effectiveness of an enterprise. The culture and technology of the enterprise are also important. Culture is manifested by the informal organization through working relationships and shared values (Allaire and Firsirotu, 1984; Bate, 1984). Technology is defined by the

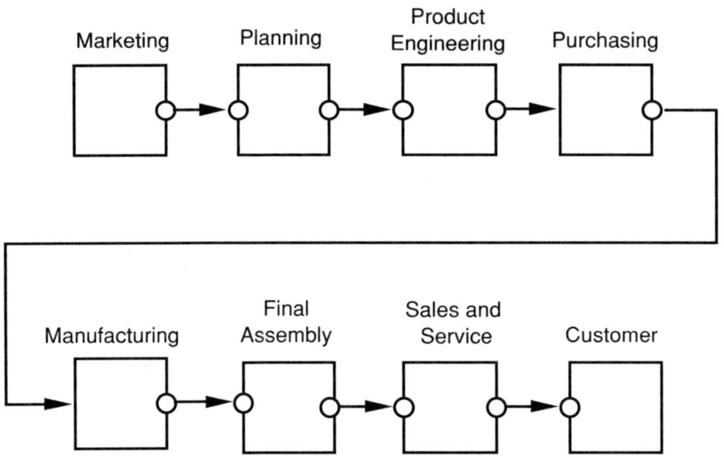

FIGURE 1 The functional organization shown with a sequential flow of work.

skills, tools, and methodologies employed by the enterprise in transforming input to output (Passmore, 1988). Thus, if we judge best structure from comparative studies, we have the problem of having to factor out the cultural and technological contributions to effectiveness, which is not straightforward to do.

Another approach, which avoids this difficulty, is to draw upon classical administrative criteria. Administrative theory, whose origins date from Taylor's (1911) publication of *The Principles of Scientific Management*, was an attempt to illuminate rigorous principles for creating successful organizations. However, the classical administrative criteria do not, as originally intended, hold the stature of fundamental principles (Simon, 1976, pp. 20–36). Simon argued instead that these so-called principles of organization are but *part of the criteria* for describing and diagnosing administrative situations. The purpose of this chapter is to identify the key criteria for diagnosing the administrative situation posed by a manufacturing enterprise and to use them to arrive at an organizational structure that should overcome the problems of the functional organization.

CRITERIA RELEVANT TO ORGANIZATIONAL STRUCTURE

The traditional bottom-line metrics for rating the effectiveness of a manufacturing enterprise are return on investment and market share. However, there are other metrics, or measures, of enterprise effectiveness, such as quality, cost, lead time (including flexibility), and innovation (Leong et al., 1990). Manufacturing enterprises that are successful in competing in highly competitive world markets have been found to score well on these metrics (Clark and Fujimoto, 1989b; Dertouzos et al., 1989; Hayes et al., 1988; and Stalk and Hout, 1990). It follows, therefore, in competitive markets, that a manufacturing enterprise should focus its energies not directly on the dependent bottom-line metrics but instead on goals that challenge the enterprise to score highly on a set of fundamental metrics that the enterprise can directly influence and which, in turn, drive the bottom line. These goals represent important criteria to be considered in arriving at an organizational structure. Another important consideration is the work plan that the enterprise uses for achieving the goals.

Thus, the effectiveness of a proposed organizational structure can be evaluated from an administrative viewpoint by seeing how well it serves the classical criteria in administering the goals and the work plan. This can be determined using a process defined by the following steps: (1) Draw the structure of the proposed organization, starting with boxes showing all the vice presidents and their titles. (2) Map onto this structure the locations where authority and responsibility for the product lies in terms of the goals and the steps of the work plan. (3) Judge whether the goals and the work plan authority and responsibility mappings pass or fail the classical administrative criteria. (4) Repeat the process until you find a satisfactory coarse-grained organization. (5) Repeat this process at the next lower level in the organization to evaluate the structure at finer and finer levels. (6) Stop the process when it begins to give bad answers.

CLASSICAL ADMINISTRATIVE THEORY

The classical administrative theorists did not arrive at one exact set of criteria. We will use those attributed to Urwick (Brech, 1958, pp. 371-378):

Functionalization: *The necessary units of activity involved in the object of the enterprise should be analyzed, subdivided, and arranged in logical groups in such a way as to secure by specialization the greatest results from individual and combined effort.*

Correspondence: *Authority and responsibility must be coterminous, coequal, and defined.*

Initiative: *The form of the organization must be such as to secure from each individual the maximum initiative of which he is capable.*

Coordination: *The specialized conduct of activities necessitates arrangements for the systematic interrelating of those activities so as to secure economy of operation. Reference from one activity to another should always take the shortest possible line.*

Continuity: *The structure for the organization should be such as to provide not only for the activities immediately necessary to secure the ob-*

ject of the enterprise, but for the continuation of such activities for the full period of operation contemplated in the establishment of the enterprise. This involves a continuous supply of the necessary personnel and arrangements for the systematic improvement of every aspect of the operation.

The initiative criterion supports modern ideas of employee involvement such as quality circles and participative management. The continuity criterion's call for "systematic improvement" is no less than today's call for "continuous improvement." Thus, Urwick's administrative criteria are very much with the times!

CHOICE OF FUNDAMENTAL METRICS

Other authors in this volume identify various metrics. Some of these—such as part counts and materials used—are very detailed, while others such as time are more general. The goals defined in terms of the fundamental metrics for an organization should be few in number and the most fundamental in the sense that they should include as a subset most if not all other more detailed metrics that are important in the fine-grained, operational structure of the organization. For example, a targeted level of quality may be an enterprise goal that includes the goal of a specific factory defined by a detailed metric of low variance from specification for the component made by the factory.

The fundamental metrics we will use here are cost (C), value-to-the-customer (V), and the pace of innovation ($1/dt$), where (dt) is the time between innovative product introductions into the marketplace. The rationale for selecting these as fundamental metrics was derived elsewhere (Cook and DeVor, 1991) based on a simple market model that yielded equations that expressed return on investment and market share in terms of these quantities. Perhaps somewhat surprisingly, quality, defined as the net value of the product to society, was found to be a bottom-line metric depending on the square of the difference between value and cost, much like return on investment. This new definition of quality is an inversion and extension of Taguchi's definition of quality loss as the loss to society as a result of a product's variance from ideal specification (see Taguchi and Wu, 1980).

TAGUCHI'S PARADIGM

A simple but powerful paradigm for the general work plan of a manufacturing enterprise has also been put forth by Taguchi (see Taguchi and Wu, 1980). This is a multistep process as shown in Figure 2. System design generates the product specifications based on customer needs tempered by practical design and manufacturing considerations. Parameter design minimizes variance in the product specifications or target values re-

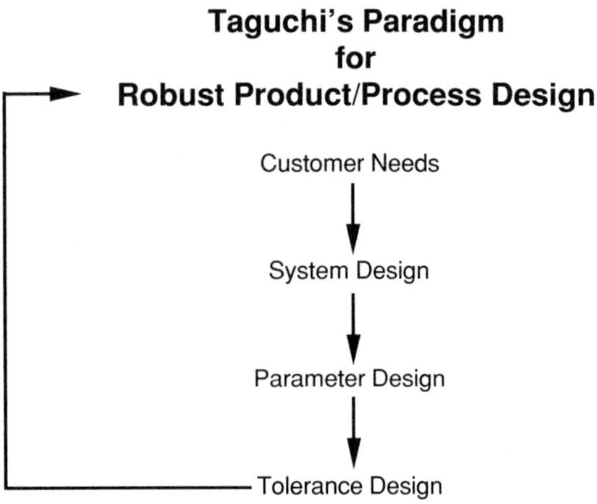

FIGURE 2 Taguchi's paradigm for achieving a robust product and process design.

sulting from process and environmental "noise." This step is based on Taguchi's definition of quality loss as the loss to society resulting from variations from specification. The final step is tolerance design, which offsets the singular nature of the added costs to manufacture a product to absolute precision by permitting a bounded level of statistical variance. Taguchi's methodology is chosen here under the strong belief that, other things being equal, products developed according to this paradigm will result in the least loss of quality due to variance from specification and also meet cost constraints. Moreover, starting with a given baseline process and product, application of Taguchi methods often simultaneously improves quality and reduce costs (American Supplier Institute, 1989).

APPLICATION OF THE CRITERIA

The statement of our problem then is as follows: "Find the organizational structure that best satisfies the administrative criteria when the enterprise, using Taguchi's paradigm as a work plan, desires to score highly in terms of the fundamental metrics." The first administrative criterion listed (functionalization) does not have to be considered any further as it is no more than a restatement of the problem. We will also hold judgment on the initiative and continuity criteria until we have looked at two different organizational structures for comparison.

It is useful first to establish a baseline by evaluating the correspondence and coordination criteria for the functional organization. The results have been tabulated on a pass/fail basis in Table 1 based on mapping the three fundamental metrics and the three steps for Taguchi's paradigm onto the organizational structure of Figure 1. The correspondence criterion fails against every objective but one because the responsibility and authority for value to the customer (V), cost (C), the pace of innovation ($1/dt$), and the Taguchi paradigm are so diffuse across the functional structure. The pace of innovation for the functional organization is given a passing grade for its "quantum leaps" but a failing grade for continuous improvement.

The coordination criterion for the product is also compromised in the functional organization as the reference from one activity to another—from engineering to manufacturing, from manufacturing to purchasing, etc.—requires crossing divisional boundaries and thus does not take the shortest possible line, which would be intradepartmental or intradivisional. The systems design step of the Taguchi paradigm requires close coordination between marketing, planning, and systems engineering. Parameter and tolerance design require close coordination between design engineering and manufacturing. All are in different divisions in the functional organization. Innovation again receives both pass and fail marks because quantum leaps often do not require much if any coordination for the initial inspiration; whereas, continuous improvement requires much coordination on a day-to-day basis. The above results would not change if the metrics used were quality, lead time, cost, and innovation.

As noted above, the functional organization's fatal flaw is that it supported a sequential product-realization process. Based on our evaluation of this structure using the classical administrative criteria, we are able to give a more insightful description of its shortcomings: Simply stated, the functional organization is unable to administer the metrics and the work plan required to face off against world-class manufacturers in highly competitive

TABLE 1 Pass/Fail Analysis of the Correspondence and Coordination Administrative Criteria for the Functional Organization versus the Three Metrics and Taguchi's Paradigm

	The Three Metrics			Taguchi's Paradigm For Robust Design		
	Value	Cost	Pace	System	Parameter	Tolerance
Correspondence	F	F	P/F	F	F	F
Coordination	F	F	P/F	F	F	F

global markets. Thus, for success in today's environment, it is necessary to arrive at an organization that groups the right elements of the product-realization process into units such that the correspondence and coordination criteria are simultaneously satisfied for the product (the fundamental metrics) and the process (the steps in Taguchi's paradigm).

There are myriad other ways than functional to divide the manufacturing enterprise for organizational purposes. When studying product design, the design process, and the manufacturing process, a useful division is often achieved by exploding the total system into subsystems which in turn factor into sub-subsystems that eventually factor into components. In what follows, we will use the natural system/subsystem (SYS/SS) architectural breakdown of a product as the proposed organizational structure for developing and manufacturing the product. We will staff the organization and place authority and responsibility in a manner that satisfies the administrative criteria when tested against the fundamental metrics and Taguchi's paradigm.

This organizational structure is shown in Figure 3. Materials are input on the left side of each box, and all output leaves from the right. Connection points for material flow are shown with a mesh fill and connection points for information flow are shown with wavy fill. Control information enters at the top of each box. Like the traditional product organization, the SYS/SS structure is an example, and a very robust example, of a self-contained organization (Galbraith, 1977, p. 51; March and Simon, 1958, p. 29; Walker and Lorsch, 1970) that should generate a strong customer orientation.

The customer is the unit labeled C. The SYS unit has the authority and responsibility for the traditional functions that consider the product as a complete unit. In the functional organization, these would be marketing, sales, design, systems engineering, packaging, and final assembly. It is mandatory, for example, that packaging and final assembly responsibility and authority reside together in the SYS unit to satisfy the administrative criteria. The chief responsibility of the SYS unit is to understand the customer's changing needs and translate them into a set of specifications for each individual subsystem. The systems unit must, therefore, have the authority and skills to achieve this important task. Its hallmark should be the ability to be close to the customer and translate customer needs into product specifications that result in the optimum balance of product value and cost for the intended market segment. It should be hard to discern, for example, any boundaries within the SYS unit between what has traditionally been marketing and systems engineering.

The units labeled SS1, SS2, and SS3 represent subsystem units. The SS3 unit is cross-hatched to indicate that it is (or could be) a supplier and not part of the parent organization. Subsystem units have the responsibility to supply the systems unit with subsystems that are ready for final assembly

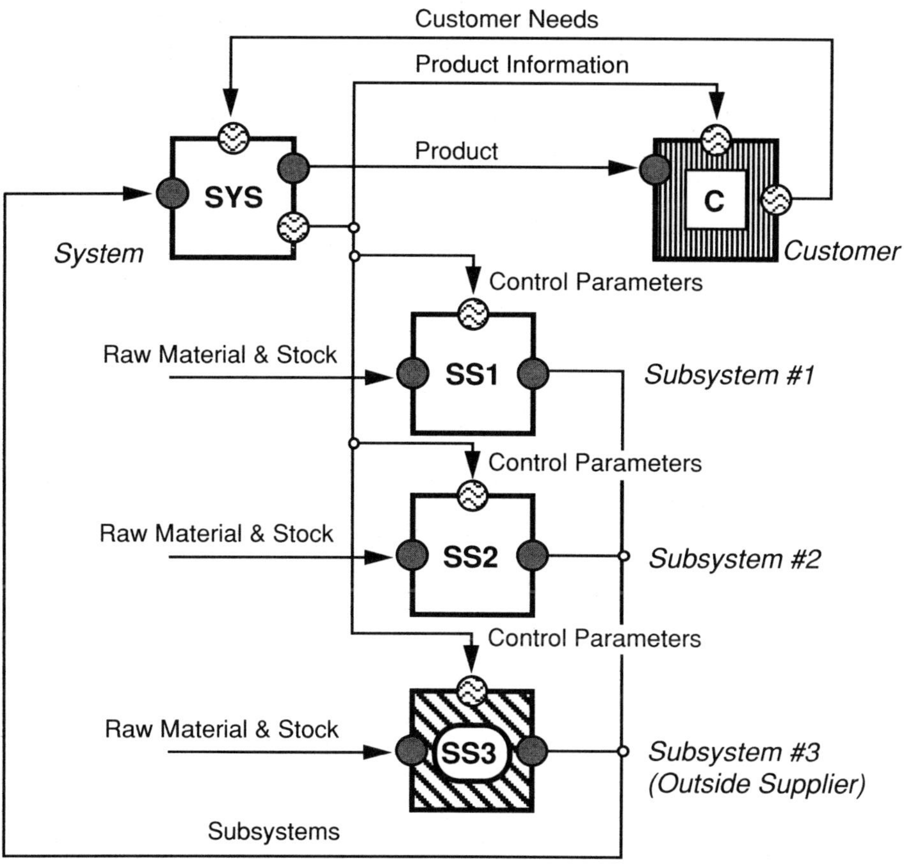

FIGURE 3 Input, output, and control for an organization based on a SYS/SS architecture. For clarity, this figure does not show direct lines of communication from the SS units to the customer or the myriad feedback channels between the SYS and SS units.

and that satisfy the system specifications for the subsystem. The latter are defined through controls issued by the systems unit. It is important to understand the basic purpose of controls; they exist so that the system is optimized for the customer. Without controls, each subsystem unit would tend to optimize the subsystem for which it is responsible, with the result that the overall system is suboptimal.

The controls include cost, weight, package dimensions, and high-level performance criteria for each subsystem. The controls should represent the truly minimal number of specifications needed to generate the desired level of customer satisfaction while leaving the subsystem units considerable latitude in arriving at the best subsystem design solution under the control constraints. The establishment of control parameters at the outset of the product realization process would likely be aided, in true simultaneous engineering fashion, by a small transient team made up of specialists from the system and subsystem units. This would ensure that the control parameters are viewed by the subsystem units as a handshake instead of a handoff.

On receiving the control parameters, the subsystem units can actively begin the product realization process. The desired result is a subsystem design that couples the product and process parameters in a manner that minimizes variability in the product through Taguchi's parameter and tolerance design steps. To achieve this result requires close coordination between those responsible for designing and manufacturing the subsystem and those in purchasing who support them. All of these persons must reside within each subsystem unit to satisfy the administrative criteria for the fundamental metrics and for the parameter and tolerance design steps in Taguchi's paradigm.

When the SYS/SS structure is staffed in the fully self-contained manner described above, this organizational structure passes (at the coarse-grained level shown) all the correspondence and coordination criteria for Taguchi's paradigm and for all the metrics except for, perhaps, the "quantum leap" form of innovation. A separate research division may be required for quantum leaps. (This same requirement probably also holds true for the functional organization.) In contrast with the functional organization, authority and responsibility are coterminous, coequal, and clearly defined for the SYS/SS organization as described here.

Both the functional and the SYS/SS structures should be equivalent for the initiative criteria. However, for the continuity criteria, they differ considerably when it comes to developing people who have the business experience to lead a manufacturing enterprise. The functional organization generally serves up leaders with narrow expertise. The SYS/SS structure, however, should give early and broad experience through routine immersion in all business disciplines. A natural consequence of this structure is that the leaders who emerge should have a full, accurate understanding of the product-realization process.

DISCUSSION

To illustrate the difference between the functional and the SYS/SS structures, consider the purchase of an automobile. The customer wants a durable, responsive engine and a reliable, smooth transmission, for example. The responsibility for value to the customer of these subsystems, however, is shared between engineering, manufacturing, and purchasing for functionally organized automotive companies. It is not coterminous at any practical level as the total responsibility for these three operating divisions generally resides with the president or chief operating officer. Each group has sufficient responsibility to become involved in every issue regarding value to the customer for these two subsystems, making it difficult at best to administer the issues. Agreements come slowly. When problems arise, finger pointing is the norm. When everyone is partly responsible, no one is. Intense competition, which causes customer satisfaction and lead time to become crucial metrics, has challenged the responsiveness of functional organizations, and many are struggling as a result.

For our automotive example, the system unit would be responsible for understanding and specifying the vehicle value-to-price relationship important to the customers within the targeted market segment. This would include specifications for vehicle style, weight, cost, features, options, and overall performance factors such as fuel economy, acceleration, ride, and handling. There is no a priori need for separate marketing, planning, and systems engineering groups with this structure. *Strict application of the correspondence and coordination criteria also places vehicle assembly operations within the SYS unit for an automotive company.*

Each subsystem unit would develop, manufacture, and assemble its subsystem in accordance with the pertinent control parameters. The latter of course have to represent what is possible according to the state of the technology, which means that excellence in systems engineering, as noted earlier, would need to be a strong point of the SYS unit. Each subsystem unit would also require systems expertise in transferring the control parameters into meaningful design direction for each of the component groups that make up the subsystem unit. Thus, the subsystem/component structure parallels the SYS/SS structure. This is shown schematically in Figure 4 for subsystem SS1, which is factored into three component units, CM1, CM2, and CM3 (outside purchased), and a unit labeled SS1:SYS. The last of these has the systems responsibility for SS1, which is, of course, a system included in but subordinate to the full product system. Most large subsystems will not factor into components immediately as shown in Figure 4 but will factor first into a set of less complex subsystems.

The SYS/SS architecture shown in Figures 3 and 4 creates a waterfall of self-contained teams, minimizing the need for transactions between units, whereas the functional organization requires many more interdivisional transactions for successful product realization. Moreover, divisions in functional organizations are different culturally, which makes interdivisional transactions difficult. The "throw-it-over-the-wall" syndrome for product realization most likely arose from the desire to minimize face-to-face interactions between functional divisions after transactions had become too tedious and adversarial as cultural differences became large and entrenched over time. *The sharp differences in operational responsibility between divi-*

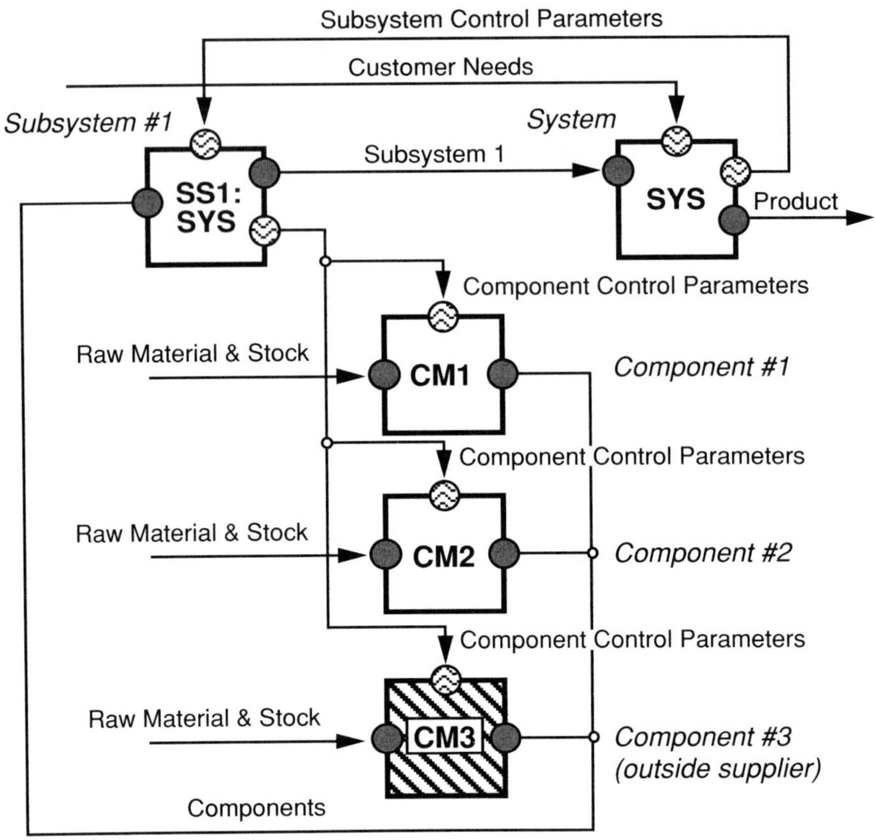

FIGURE 4 Input, output, and control at the subsystem and component level.

sions in the functional organization are most likely the root cause for their sharp cultural differences. By contrast, units in the SYS/SS architecture should have similar cultures, because on average each has similar types of responsibility, including engineering, manufacturing, purchasing, and financial control. Thus, the SYS/SS structure should achieve objectives more quickly because fewer transactions are required between divisions and because those that are needed should proceed smoothly because the cultural differences should be small (Cook, 1991).

Although the emerging practice of simultaneous engineering should improve the performance of functional organizations, the results will likely be suboptimal because team members will often have divided loyalties between their parent functional divisions and the team. It is also difficult to transfer responsibility fully for the value, cost, and the pace of innovation away from the operational divisions and to the teams, because the teams generally have a finite lifetime. The divisions, on the other hand, have an indefinite lifetime and thus have to live with the results that the teams produce. Present efforts at simultaneous or concurrent engineering may be a classic attack on symptoms, resulting in a suboptimal "band-aid" and not a solution to the root cause of the shortcomings of functional organizations in today's environment. Moreover, since the functional organization is likely to be the source of the cultural differences that impede the effectiveness of an enterprise organized along these lines, these cultural differences should persist as long as the source for them remains.

The SYS/SS architecture, however, is a much more robust formulation of the team concept than simultaneous engineering by creating teams at every level and by coupling more strongly actual production plants to the design process. In total, these actions should bring the organizational structure into better harmony with the requirements of the Taguchi paradigm and classical administrative criteria for value, cost, and the pace of innovation.

At some point in unfolding the organization below the level shown in Figure 4, the arguments used to develop the SYS/SS structure could begin to generate specialists with not enough work to do. This would define the point where some traditional functional responsibilities would begin to appear in the organizational structure as opposed to full system, subsystem, and component responsibilities. However, if people can handle more skills than they are given credit for today in most manufacturing enterprises, then the SYS/SS structure could extend to a finer level of the organization. There is great discourse about wanting to reduce the number of job classifications on the plant floor. The proliferation of specialists in the office should also be challenged (Hayes and Jaikumar, 1988). Marketing and systems engineering skills are closer together in the scheme of things according to both the Taguchi paradigm and the quality function deployment (QFD) process (Hauser and Clausing, 1988) than one might expect from the separate way in which these specialists are formally trained. The same can be said of component design engineers and component manufacturing engineers.

Customer Satisfaction

HAROLD E. EDMONDSON

The phrase "customer satisfaction" seems to appear in print more frequently than any other catch phrase used to describe a new-found magic for industrial success. This phrase not only is used by most business authors, but also finds its way into virtually every annual report and statement of business strategy published by companies, large and small, across our land. I do not want to downplay the importance of customer satisfaction. In fact, I believe it is the cornerstone of our rebirth as an industrial nation. I do, however, feel that a thorough understanding of this topic is missing from the minds of most of the people that claim to be practicing proponents.

To help us to think clearly about this complicated topic, I would like to propose several important elements of customer satisfaction that need careful thought. Reading this chapter certainly will not make you an expert on customer satisfaction, but I hope at least to highlight some of the pitfalls. Clearly, other readings would be helpful. The difference between providing customer satisfaction expertly and doing it poorly is frequently the difference between success or failure of your company.

In the ensuing paragraphs when I refer to customer satisfaction or to customers in general, I am including two large bodies of people. The first falls under the traditional description of a customer, that is, those people outside of your firm who buy your service or products and reward you with money in return. One very important task, and one we will assume you have already done, is to segment your market carefully. In other words, I assume that you have defined your marketplace and properly understand the differentiation between the people you view as customers and those you view as being outside of your intended marketplace.

The second set of people whom I view as customers, but who are not always given this revered title, are those departments within your company who are served by your department or yourself. It is entirely appropriate (virtually necessary) that every suborganization within a company view the recipients of their work with as much awe and reverence as you do their outside customers. This set of internal customers is probably hardest of all to satisfy in that fairly often departments within the same firm or division take each other for granted. The assumption is erroneously made that everybody knows what everybody else needs and, by virtue of having the same name on the paycheck, all departments will work in glorious harmony. This is seldom true.

This lack of goal congruence between departments of the same firm can have its roots in a variety of cultural elements. Sometimes teamwork is not practiced as well as it should be. Sometimes capabilities and expectations are not in concert. But most often, I feel, departments do not worry about the needs of their partners simply out of ignorance. I think, in general, people want to do as good a job as they can, but often it simply does not occur to them to walk across the aisle and ask how well their department meets the needs of their partner departments. Meeting the needs of these internal customers is just as important as concentrating on your outside customer needs, and most of the comments made in the ensuing paragraphs will address both types of customers.

WHO REALLY IS THE CUSTOMER?

The question of defining who your customers are seems fairly easy, particularly if you have segmented your market properly and understand who you are trying to satisfy. However, a subtlety that frequently goes undetected by many firms is that the customer set can be divided into two parts—the apparent customer and the user. The apparent customer is the person or group of people who decide what product to buy and basically have control over the purse strings. The user is the person or group who physically uses the product or is the direct recipient of a service.

One way to illustrate this significant difference is to relate a new product story from the history of a well-known dog food manufacturer. After surveying their customers carefully, the manufacturer decided that the greatest unfulfilled need their customers had was to somehow provide a food that overcame the greatest disadvantage of man's best friend, specifically, the dog's bad breath. It did not take the research department long to come up with a chlorophyll additive to their regular product, and they went into production quickly. The first several weeks after the product introduction were quite successful for the firm in that they sold lots of their product. They smilingly congratulated themselves for having accurately ascertained their customers' needs and then satisfying that need. Unfortunately, the story does not have a happy ending in that the repeat sales of this product were virtually zero and rather shortly the product turned into a dismal failure. The simple reason; the dogs would not eat the product. Remarkably, the firm put all their bets on understanding the customer and spent no time understanding the user. The point of all this is that, to truly capture the fancy of your customer you must recognize that both the apparent customer and the user must be satisfied. Forgetting the apparent customer will probably preclude your making even the first sale, and forgetting the user will undoubtedly, as with the dog food company, preclude your getting any follow-up sales.

WHAT DOES "SATISFACTION" REALLY MEAN?

As in defining "customer" above, defining "satisfaction" also appears simple. However, as with "customer" there is a subtlety that needs addressing. Satisfaction, by most definitions, simply means meeting the customer's requirements. However, these requirements frequently fall into two categories—needs and wants. I would define needs usually as the real requirement to which the customer should be putting the product, or, the true requirement of the customer. Wants, on the other hand, are the perceived needs the customer feels should be met. If you are lucky, the customer's needs and wants are synonymous. But this is seldom true in real life.

The situation is further complicated by the multiplicity of features that a customer looks for. Seldom is a buying decision made on a single feature of a product. More often several factors, both needs and wants, influence the decision.

There are a number of ways in which this complicated wants-versus-needs syndrome can cause problems. Three of the most troublesome of these illustrate the difficulty involved in sorting out this part of your product definition.

The first is the customer who knows what he wants but does not know what he needs. This scenario is particularly troublesome if the customer thinks that his wants are, in fact, actual needs. Getting to the core of what your customer's real requirements are takes a great deal of patience and skill, but it is obviously imperative that you pursue this process until you are absolutely certain you understand the customer's situation.

The second difficulty is the customer who has a hidden want but purposely masks it as a need. This customer may have an affinity for a competitor's product and chooses an obscure feature from one of your competitor's products and labels it as an absolute need with the hope that you will go away and not bother him since you do not have a similar feature. This customer may be so enamored with your competitor's product that you never will make a sale, but it is vitally important that you not chase phantom specs only to spend your money on something the customer really does not need.

The third pitfall in distinguishing wants from needs is finding a customer who is all too willing to tell you what your next-generation product should be in terms of your traditional product. You will be miles ahead if you can discipline yourself to concentrate on understanding what customers plan to accomplish with your product, as expressed in their terms. Do not let them describe your product to you by using your specifications; this certainly will deprive you of breakthrough opportunities. If you are creatively trying to construct a product to meet a customer goal, you need to start with an open mind, not one cluttered by thoughts of simply extending the features of an existing product. It is true that customers certainly can be creative, but, in general, that creativity is diminished dramatically when they describe their solution in terms of your traditional product specifications.

There probably is no simple way to differentiate between your customers' needs and wants, but one helpful way to look at the problem is to rephrase your objective; set your sights on helping your customers meet their goals. This rather concise objective should require you to understand both their needs and their wants and, further, nudge you into a much more open-minded approach toward providing truly imaginative products.

Sometimes it may appear that "wants" are not at all what the customer needs and are put on this earth only to deceive well-meaning product managers. Whether that is true or not I will not debate. However, I will state unequivocally that at the time the customers state their needs, whether they are needs or wants does not matter much. They are both perceived to be needs in the eyes of the customers and must be dealt with either by providing them to the customer or, in some case where they are very far-fetched, by persuading the customer that there is a better way to meet their goals.

WHAT DO YOU WANT OUT OF IT?

Somewhere along the line, before getting down to your actual product definition, it seems appropriate to sit back and consider what your objectives are in this endeavor to satisfy your customer. Are you interested in short-term sales, or are you interested in a longer-term relationship, or both? Do you want to make high profits on a few sales or lower profits in a larger marketplace? Are you interested in providing the customer with some value that you may not get paid for in the short term but, through a closer customer partnership, will earn your return over a longer period? It is not my intention to suggest which of these is best for you or that any of the possible approaches is somehow better business or even more ethical business conduct than the others. I am simply suggesting that it is important for you to understand yourself as well as it is for you to understand your customer.

At Hewlett-Packard we prefer to aim our relationships with our customers toward the long term. Generating a customer's loyalty is probably the single most important goal we can hope to achieve. Certainly we are interested in growth and profit and market share, but we feel that building customer loyalty is probably the best way to accomplish all of our objectives.

THE CRITICAL STEP—THE PRODUCT PLAN

By far the most critical step, in my opinion, on the way toward customer satisfaction is the skill with which you engage in product definitions. The foundation of good product definition methodology, in HP's opinion, is the quality function deployment (QFD) approach brought to us by the Japanese and Americanized through the efforts of a number of excellent U.S. authors (see Hauser and Clausing, 1988).

For the purpose of this chapter, it is sufficient to say that the description of a product with QFD, or any process, needs to be detailed and based on a thorough understanding of the customer's requirements on a feature-by-feature basis. Additionally, the description must reflect complete understanding of what your competitors are offering today and are likely to be offering in the future—once again, on a feature-by-feature basis. Although other reading will help you understand this process, some important elements are not covered in most of the QFD texts. These elements generally are concerned with how information is obtained for product definition. One of the common failings of a firm engaging in product definition is to assign a single person or function to do the job. Many firms feel that marketing is the function that should be responsible for product definition, while others feel that the engineering department is responsible. It is my very strong view that both departments must act in partnership to gather the information necessary to form a clear, concise product definition. There is an appropriate division of duties between R&D engineering and marketing within the task of product definition. I suggest that the marketing people be held responsible for understanding the marketplace, that is, knowing who are the dominant customers and users of products similar to the one you are trying to define. Further, marketeers should have a good idea of which customers are the innovative leaders and best suited to describe future needs for the industry. Additionally, the marketing people are more likely to have insights into what competitors are about to do.

On the other hand, the engineering department is best qualified to bring technical expertise to bear on customer problems and potential solutions. Engineers also probably have a better idea about the technical suitability of the products that are currently being offered by competitors. Stated more succinctly, the marketing people should be responsible for developing the market research plan and managing the places where information is sought, while the engineers should be responsible for providing the technical creativity and working with customers' technical people to develop a set of features for the new product.

As engineering and marketing go about establishing their product definition, somewhere there exists a role for manufacturing. Depending on the products and the expected customer benefits, manufacturing can be brought in at a variety of points in the process—usually the sooner the better. This is not to suggest that manufacturing will play as critical a role as marketing and engineering, but there are quite a number of products where some manufacturing creativity can, and has, dramatically improved the customer's acceptance of a product.

The advantage in requiring that all three departments (marketing, product engineering, and manufacturing) be involved in the product definition phase is twofold. The obvious advantage is that the chance of success if much better if a task is viewed from these three distinct viewpoints rather than through only one set of eyes. The second advantage is somewhat less obvious but possibly even more important. It is that, when constructed in this fashion, the product definition becomes owned by all three functions. Any disagreement they have must be hammered out by all three functions, and the final definition is supported by all of them because their names are on the dotted line.

When dealing with inside customers, there are two facets to customer satisfaction; the first is of a technical nature and the second is of a business nature. It is conceivable that some department managers understand both facets of satisfying the customer, but frequently it is necessary to do this sort of research with a business manager (often the department head) and a technically responsible person (usually an engineer).

At this point in defining your product, it is important to remember that customer satisfaction does not stop when the customer receives the product. You have an ongoing responsibility that includes training, support, and service long after the sale has been made. These features need to be incorporated in the original product definition to be of maximum competitive impact. All too often HP and others in our industry have added service and support considerations after the product has been invented. This almost always leads to a marginal customer benefit—marginal either in its effectiveness to the customer or in its cost to you, or both.

Another overlooked potential pitfall is that the crucial elements of a good product definition analysis (i.e., customer needs and competitor offerings) can, and usually do, change dramatically with time. Thus, truths that were virtually self-evident a year ago may be completely outmoded today.

This is not to imply that you should continuously refine your list of product specifications. Once you have started designing a product (unless something catastrophic happens in the marketplace), you should severely limit the changes you make in that product. However, when it comes time for the next-generation product, you do need to reassess and revalidate all the assumptions that went into your product definition analysis.

A significant example of this phenomenon is a product that HP invented a few years ago. This product provided the customer with eight channels of input into a digital analyzer. We carefully talked to leading customers before inventing the product and were assured that, yes, eight channels were perfectly adequate. We invented the product that way and enjoyed rather substantial initial sales. After a year or two, however, we began to hear that eight channels were not what the customers wanted and that sixteen channels would be much better. When we questioned the very people who led us down the eight-channel path, they defended their position by saying that, at the time, eight channels were so dramatically better than the one channel they had been used to that it seemed perfectly adequate to them. However, after using the eight channels for a year or so they recognized that their needs were far more complex. In other words, their expectations had changed. Ironically, we had been the ones to change them but yet did not recognize the full implication of our product offering.

One last warning flag. In our view, it is important to be aware of what we think of as functional filtering. In almost any firm, consciously or unconsciously, manufacturing, R&D, or marketing is viewed as the premier or dominant function. Historically, through the 1950s and 1960s, U.S. firms tended to view product engineering (R&D) in this dominant role. We all know this led to a less than competitive set of manufacturing skills.

The Japanese were quick to reverse this and installed manufacturing as the dominant function. In my view, while this has not been as devastating as our choice of R&D, there are some signs that the Japanese are overly obsessed with manufacturing considerations. The best example of this is in the consumer electronics business, where both compact disc players and videocassette recorders have an extremely low selling price but carry with them a rather poor record of mechanical reliability. I suspect, although I cannot prove it, that this phenomenon is driven by the manufacturing people who dominate the product-definition decisions and who are horrified at the engineering department's suggestion that another $20 on the selling price might provide a much more reliable product. Thus, in a few cases in Japan, and in many cases in the United States, the dominant function has filtered good product ideas that could have been delivered unobstructed to the customer by the secondary function.

In our view, to completely satisfy the customer in the future it will be necessary for successful companies to learn that whatever will benefit the customer should be provided directly by the function best equipped to provide it and that other functions cannot be allowed to filter those good customer ideas with their own traditional thinking.

THE LAST PIECE OF THE PRODUCT PLAN—METRICS

As with any business plan, it is necessary to establish ahead of time some benchmarks to make sure that the product plan really is a good one. Obviously we could wait until the product is introduced to the market and see if it sells well—if it does, the plan was satisfactory. Obviously, sales are the ultimate test, but it seems prudent, considering the high cost of most new product programs, to set some benchmarks or tests along the way to make sure that you were right in establishing of your product definition. We suggest that there are two important times when you can take a meaningful "health check" of your program.

Both checks involve returning to your customer set (or at least the most important parts of it) as the only source of meaningful reaction to your plans. In each case, the check occurs when you have significant additional information to share with the customer. Obviously you cannot call back every five minutes and ask another question about a feature. Thus, you should limit questions to times when you really have a significant question to ask.

The first checkpoint is when you have completed your product definition. During your initial meeting with the customers, you were simply asking for features they wanted. Although there was undoubtedly some technical exchange with your engineers, the customers really have not had an opportunity to see the entire feature set put together as a complete product plan until you have concluded this phase of your work. Thus, it can be very productive to take those complete plans back to your key customers with the simple question, "Is this what you had in mind?"

The second checkpoint is when you have a good working prototype so that you can show the customer a tangible product that came from the definition arrived at earlier. The question is much the same: "Is this what you had in mind?" But this time the customer can answer with a much better idea of your intentions, and you can be much surer that you are getting a thoughtful response. All this clearly takes a lot of work and effort on your part and some imposition on the customer. Some firms might be tempted to establish metrics that they can generate and test from within the firm itself, such as meeting 80 percent of the customers' needs as shown on the QFD chart, or bettering three-fourths of their competitors' specifications. Metrics of this sort can give some indication of how well you are meeting your product definition but certainly seem to be a poor substitute for a real, live customer reaction. After all, in the final analysis the customer's reaction to your product has a nearly 100 percent correlation with whether or not your product will sell.

SANITY CHECKS

Now that we have established our plan for achieving customer loyalty, there are a couple of things that we need to revisit in order to make sure that the product plan we have just constructed has a good chance at success. The first question we need to ask ourselves is, Do we have the resources and skills necessary to carry it off? Frequently, aggressive people (whether they be in marketing, engineering, or manufacturing) tend to engage in wishful thinking when it comes to assessing how a new product program will affect market share. While I am a proponent of setting high objectives for ourselves, I think that there comes a point when the hoped-for results are unrealistic, and now is a good time to assess that.

Another thing that needs to be questioned is whether the product plan is consistent with your company's basic business strategy and charter. It may be a good plan, but, if it varies widely from your basic business, there can be far-reaching, disastrous effects in financial or sales-channel issues. I am not proposing that product plans never vary from our charter and strategy, but I am strongly suggesting that, if they do, we need to take another cut at a much broader set of issues and make sure we understand what we are letting ourselves in for.

The last sanity check is a simple question, but it needs to be asked. That is whether this product plan that we have just meticulously constructed will lead us to an overall performance that is expected by our bosses, whether they be department heads, company presidents, or boards of directors. Will this plan lead us to the proper profitability, the proper market share, proper growth rate, or whatever else has been deemed important by the people that sign our paychecks? If the answer is yes, then full speed ahead.

IMPLEMENTATION—THE EASY STEP?

Now that we have constructed our product plan, we are faced with a simple matter of implementing it. I believe that this chapter outlines the most difficult tasks facing industry today. However, I do not think that implementation is easy by any stretch of the imagination. Implementation is certainly difficult and has been addressed aptly by the authors in other chapters of this book.

The Interface Between Manufacturing Executives and Wall Street Visitors—Why Security Analysts Ask Some of the Questions That They Do

PHILIP A. FISHER

Many in industrial corporations who do not have close association with the investment community do not realize that those interested in selecting which stocks to purchase come from two rather unlike groups with quite different fundamental objectives. One group feels that the intelligent way to handle stock purchases is to study the various factors that should cause a particular stock to rise in the near future, buy it before that rise has gone very far, sell it when the rise has run its course, and then look around for another vehicle or even the same one to repeat this happy set of circumstances again and again. The time horizon of such stock buyers is relatively short. Their period of holding runs anywhere from a few weeks to a year or two. Such holders are essentially short-range in their goals. The other group, which under the psychology that exists today is much smaller but probably has total holdings at least comparable in size to the short-term group, looks upon stock ownership in a very different way. Investors in this group feel that in today's inflationary world ownership of unusually well-run corporations that are both low-cost operators in their industries and offer strong possibilities of growth is an excellent way to store wealth. Such investors will remain stockholders in the companies they choose until they see a fundamental change in the characteristics of the particular company.

No accurate statistics are available as to the value of the ratio of total outstanding shares owned by either group versus the other. Both are very large. However, from the standpoint of the number of transactions occurring in any given period, the short-range goals of the first group imply that those buyers will initiate many times more transactions than the long-range holders. Therefore, it is only logical that those who represent the large number of brokerage houses that earn most of their income from commissions on completed transactions, together with the many other investment management firms receiving counseling fees for advising or managing funds desiring this type of management, result in the majority of Wall Street type visitors to industrial corporations asking questions that are quite short-range oriented.

Short-range buyers are not nearly as interested in the fundamentals of the business. The element that causes a stock to rise or fall in the immediate future is the general realization of some significant favorable or unfavorable factor not previously recognized by the general market and therefore "not in the price of the stock" until it does become recognized. Nevertheless, there is less risk in buying stock for a near-term gain if the fundamentals are really secure. This means that an intelligent analyst even with short-range goals will pay some attention (but usually not much) to the fundamental strength of the company. Similarly, even the most long-range potential investor wants to purchase at a time when a stock is not at a short-range peak or facing immediate bad news. Therefore, such long-range holders will also pay attention to the immediate future but give it less weight.

In my personal view, for management to give any more attention than necessary to short-range buyers makes about as much sense as it would for a construction company to use blocks of ice to build a bridge across a river in the tropics. Short-term holders surge in and out of a company's shares, usually doing so at just the time when it will accentuate the price fluctuations that are inherent in any widely traded stock. This works against the corporate good. It detracts from the feeling of both key people and all employees that long-term holding of company stock is of real benefit to them. It deceases the value of stock ownership as a tool for recruiting key people and improving the morale of all employees. In this way, short-term trading erodes the loyalty and productivity of a company's employees, at just the time when these qualities are most needed, because accentuated declines in the value of their own holdings will accentuate employees' fear and lack of confidence.

There are several reasons why top corporate management gives as much time to some of those whose interest is purely short range as they do. They may not know how to recognize from which group a particular visitor comes. They also may be dazzled by some of the "big names" of short range visitors or even more by the large size of purchases that such visitors frequently are prepared to make, with the resulting increase in the market price of the shares. They may be concerned that if they do not give as much time to one investment representative as another, such a policy may very well be frowned on by the Securities and Exchange Commission. The irony here is that speculators, because of the nature of their interest, are more likely to ask the kind of questions that, if answered, would be considered inside information because it might have an immediate effect on the price of the shares.

Most sophisticated financial people know the rules and, for example, for some weeks before a quarterly statement is to be issued to the public, will not ask company officials how the sales or earnings for that period have gone. Similarly, they will shy away from asking about the receipt of especially large orders or contracts that are very big in relation to the size of the company. However, with their basic objective of a near-term increase or decrease in stock prices, which nearly always change when something is revealed that is contrary to the prior thinking of the financial community, short-range investors are overwhelmingly dedicated to finding out such information if they can do so by indirect means. This is why they ask about such matters as total production capacity and the percentage of capacity currently in use, because the subsequent discussion may well throw significant light on near-term trends.

Why are the questions of long-term investors quite different? It certainly is not because long-term investors are less eager to make big money right now, that is, this month or this year, than anyone else. Rather it is because many decades of experience have shown that those who are constantly trying to profit by getting in and out of stocks over short periods are playing a game that is almost completely unwinnable. The factors affecting a short-term change in stock prices are so complex and cover so many influences beyond the quality of the company itself that they are not subject to a consistent source of gain, regardless of the training, brainpower, or computer power of the speculators. I would suggest that those tempted by what looks like a logical means of increasing wealth look around and see the number of people who have amassed real wealth in this way compared to those who have done it by finding unusual companies and sticking with them for a long period of years and until such time as a fundamental change in the character of the particular company may occur.

Appraising the manufacturing competence of an industrial company is a significant factor in judging not only the important contribution that the manufacturing arm can give to the total success of the company but also an excellent way of judging the competence of top management itself. To understand this point, it is necessary to go back to basic business fundamentals. What must a company do to become a successful long-term investment? A company must make a better product less expensively than the competition and provide a better service.

It is because the importance of the manufacturing arm of a company is so great in contributing to superior total overall results that Wall Street visitors to key manufacturing executives ask some of the questions they do. One question almost sure to come up concerns quality. What has been the rate of improvement in eliminating defects, expressed sometimes as improvement in yield? How has it been brought about, and what further improvements are being planned? Superiority in this area is a double winner. "Making it right the first time" causes costs to go way down. Simultaneously, causing customers to feel that products shipped to them can be depended on in a way that competitors' products cannot, nearly always results in increased market share. Through economies of scale, these increases in market share usually cause costs to go down still more with a further increase in profit margins.

However, these happy results do not come of their own accord. Neither will they continue unless the conditions that bring them about are carefully nurtured by management. Therefore, the bulk of the questions from a financial visitor primarily interested in truly great long-term growth will center on the management's actions that initiate these conditions and then ensure that they continue. These matters overwhelmingly are involved with the handling of people. Contrary to popular belief, capital investment, while helpful, usually is less significant than "people leadership." This is because no matter how brilliant the top management of a company may be or how high the IQ of its industrial engineers, it has been shown time and time again that these brilliant people peering down from above do not see ways of making operations more efficient that can be pointed out by some people who are in the midst of doing these things themselves. Furthermore, if small teams of people have confidence in the overall fairness of their company and are permitted to set their own goals and, even more important, have confidence that they will be personally awarded a fair share of the savings to the company that better productivity brings to the earning statement, production will rise to levels far higher than it would seem reasonable to attain by any yardstick set from above. The climate for this type of cooperation cannot be achieved by the manufacturing executives alone although they may play a very significant role in accentuating a favorable atmosphere created by higher authority. Awareness of these factors can cause a financial visitor to have far more interest in the answers to some of the questions he asks manufacturing personnel than that of just judging the abilities of such personnel themselves. The answers can have great significance as to the quality of the company as a whole.

These questions of good "people relationships" go far beyond those I have just covered which essentially involve the relationship of blue-collar workers with their immediate bosses. Great as are the variations between one company and another in the degree of productivity from this area, the variations are probably even greater between companies in the unbelievable costs that arise when there is lack of real cooperation between manufacturing and research. When research or engineering simply designs a product and then "throws it over the wall" and says to manufacturing, "Now, go make it" the result usually is that countless dollars and even more valuable time is wasted until the two groups work out a method whereby a product is manufacturable at a competitive price. Furthermore there is the even greater danger that, in the time delays that this causes, a competitor will get to market first and have all the advantages that being there first usually brings with it. A similar tale of misadventure can occur if the marketing arm of the company is not in close touch with the various stages of inventing and making the new product. Again revisions may then have to be made to satisfy customer needs. This wastes still more money by "having to do it twice." It can open up time to market advantages to a competitor who has made it right the first time. Unfortunately, in too many instance these failures of real cooperation do not just exist between the various major divisions of the company but also occur within manufacturing itself. This is when parts or components are made by one plant under one manager and then shipped to another division which uses them in their assembly. Human pride and the NIH factor are so deep within most individuals that it takes a truly superior overall management and an equally superior manufacturing staff to open up the advantages which a really well-run company can attain by getting genuine cooperation between all phases of a company's activities.

If a financial visitor asks some questions pertaining to these matters in too direct a fashion he may receive a pleasing answer rather than a factual one. This is why some of the questions posed to manufacturing people may be asked for reasons other than what may appear on the surface and may produce some of the puzzlement among manufacturing executives of why they are asked some of the questions they are.

What are some of the results that can be obtained by the right management climate and a superior motivation of people other than the superior yield and high output per worker that I have already mentioned? One is a sharp reduction in the production cycle time, that is the period from the time raw materials are left at the receiving dock to when product is shipped to a customer. Savings here, aside from the obvious one of freeing working capital through smaller inventory requirements, are numerous. The customer can usually get faster delivery of a new product. There is also less danger from theft about which nobody likes to talk but which is highly costly to most businesses since the products are that much less exposed to the danger of thievery. The cost of tool cribs and the workers employed therein are reduced since it is unnecessary to lock up as much each night.

When continuous progress has been made in matters like these something else has been attained that I believe is even more important. This is that the average person working in all types of positions throughout such a company feels that just not the heads of manufacturing but top officers as a whole are really competent and KNOW WHAT THEY ARE DOING. All levels of personnel will listen and cooperate when some new and different way of doing things is presented to them by their bosses. Just a decade ago such techniques as have proven so successful as statistical quality control and shortening cycle time were virtually unknown. Today, their results are recognized. The competitive pace of the 1990s will surely produce more and different ways of making major competitive improvements. In some companies most personnel recognize that this type of innovation has been both successful and has benefitted them personally. They will be ready to change their ways and cooperate in some other and today undreamed of methods of still further improving efficiency that are sure to show up in this new decade. It will be very much more difficult to bring about such advances in other companies where this type of basic respect for top management has not occurred. These are not easy matters for a financial visitor to obtain truly realistic answers. However his need of obtaining such answers can well explain why some manufacturing executives who are in the very midst of seeing these things happening are asked indirect questions the reasons for which they may not fully understand.

Following the old Chinese proverb that "a picture is worth a thousand words," the final part of this paper consists of questions of a type which financial people interested in the truly long-range prospects of a manufacturing company might well ask a V.P. of manufacturing. With each is a comment as to why the financial visitor may be asking what he does. For my own investments and those I handle for others, I am interested only in companies that recognize that competition is steadily improving, so that it is incumbent on these companies continuously to improve their own efficiency and never to be satisfied even with the quite magnificent strides that some of them have made in recent years. I try to set the same types of standards for my own work. Therefore, with this in mind and because many who may read this paper are far more expert than I am in the mechanics of some of the matters I have tried to cover, I would be greatly interested if any of such people would care to comment to me on their answers to either or both parts of Question No. 13 on the following list.

Questions That Might Be Asked (But Usually Are Not) by Security Analysts When Questioning a Senior V.P. of Manufacturing:

1. QUESTION: In regard to Statistical Quality Control, when did you start putting the first part of your operation under this technique, what percent of production is using this method now, how great are the benefits, and what further gains do you see ahead from this method?

COMMENTS: Statistical quality control is not new. It was one of the first techniques adopted by leading edge American manufacturers when it was realized drastic action was needed to survive against Japanese competition. It also is something that takes time to develop to its maximum potential. Learning when it was first started may indicate how far along the company is in this key matter. It also may furnish a clue as to whether the company will quickly take advantage of other improvements that may appear in the time ahead because an innovator or early adopter of this technique is apt also to be a leader with others.

2a. QUESTION: (For companies primarily selling to other manufacturers) How many customer awards have you received so far this year, and how many in the past two years?

COMMENTS: These awards are customarily given to the relatively few companies that are capable in two successive quarters of delivering zero percent defects and 100 percent on-time delivery. Such action usually convinces a customer that it is unnecessary for him to undergo the essentially additional expense of providing inspection on the arrival of a shipment. At the same time, it enables him to take the full advantage in inventory savings of just-in-time delivery. Such awards are usually followed by an increase in the market share of those who receive them at the expense of less efficient competitors. Consequently, they are frequently an accurate indicator that the market share of the recipient will increase.

It might interest manufacturing personnel to know of a reverse twist by which an unusually able investor relations manager of an outstanding manufacturer uses this subject. When he is first visited by a representative of any large financial institution about whom he knows very little, he deliberately brings up the subject of the vendor awards his company has received. When he finds that on talking of this subject his visitor's eyes glaze over and the visitor endeavors to change the subject to what the investment relations officer considers will be next quarter sales in one of the company's more glamorous lines, it provides a strong clue that his visitor's interest is purely short-ranged and speculative. Consequently, he tends to get rid of the visitor as quickly as diplomacy will permit. On the other hand, if the visitor shows real interest in the subject he concludes that here is a representative of a genuine investor who has a long-range interest in the company. He exerts himself in every possible way to try and provides the information his visitor seeks!

2b. QUESTION: (For manufacturing companies selling primarily into the consumer markets) Does manufacturing regularly receive information concerning ALL customer complaints that are in any way related to the nature of the manufacturing process or which are matters about which manufacturing may be able to take some remedial action?

COMMENTS: This question should immediately lead to a discussion of what actions have been taken and can be taken against complaints. A general knowledge of the shift in percentage of sales against the U.S. automobile manufacturers and in favor of those from Japan and certain European competitors should indicate to anyone the importance of pursuing this matter which at least in part is a manufacturing concern.

3. QUESTION: How early is manufacturing brought in by engineering or R&D in the development of a new product or a modification of an old one and is sales or marketing also working closely with manufacturing and engineering on these matters?

COMMENTS: This question should naturally lead into the key matter of the coordination between engineering and manufacturing not just at the initiation of a new product but all through the steps until the new product is turned over to sales. Only in this way can costly waste be avoided by making it unnecessary later to modify designs so that they can be manufactured at optimum efficiency. This is a place where rather a huge burden of waste occurs in many businesses not only through the extra cost of having to rework designs after they have first been set but through competitive losses resulting from other companies getting their product to market first.

4. QUESTION: How many quality control people do you have in your organization now, did you have three years ago, and how many six years ago?

COMMENTS: This is deliberately devised as a "trick" question so as not to get an answer that may be colored somewhat by what the person being questioned thinks the security analyst may want to hear. In recent years it has been proved conclusively that the old idea of having quality inspectors examine products either part-way through or at the end of the line is an extremely inefficient one. It soon becomes a game for the people on the line to see how smart they can be in getting products by their adversaries. The Japanese have taught us that a vastly better way is to ensure that every employee in the production process regards the person at the next step in the process as the "customer" in the same sense that another company is the customer when the product is finished. At the same time the person one step more advanced on the production line considers it his or her job to be sure the quality is right in the partially worked products that they are receiving. The savings by getting production people enthused over this sort of responsibility is quite significant and of sufficient importance to justify learning the real facts by asking a question that brings up this matter indirectly.

5. QUESTION: Exclusive of foreign plants in countries where unions are specifically required, what percent of your manufacturing employees are unionized and what percent non-union? On the same subject, how many strikes have you had anywhere around the world in the last six years and what was the reason to cause them?

COMMENTS: For the creation of truly efficient manufacturing with little waste and abnormally high yield nothing is more important than the good will and effort of all elements of the personnel. This again is a matter about which it is sometimes quite hard to get a true picture by direct question. This is but one of several questions designed to approach the matter from various angles.

6. QUESTION: How many vendors do you have today, how many did you have three years ago, and how many five years ago?

COMMENTS: Not so many years back it was considered good management to have half a dozen or more vendors bid on each job, stimulate them to compete at the lowest price possible with the thought that this will improve profit margins. Again, we have learned from the Japanese. A far more profitable long-range course is to select a very few vendors who can be depended on for superb quality and on-time delivery. Equally important, such vendors are ones who can be trusted to keep confidential long-range plans for new products so that they are both prepared and may have contributed ideas for the development of such products rather than being asked to bid at the last moment when they are not really sure of what they are doing. Therefore, a favorable answer here should show a sharply decreased number of total vendors over this time period even though the business itself may have been growing nicely.

7. QUESTION: Who are three of your vendors whom you could consider outstanding and why have you chosen them?

COMMENTS: The answer to this question can have a double benefit for the security analyst. From it he may learn of an unusually attractive potential investment about which he had previously been totally unaware. More directly affecting his immediate study of the company being questioned, frequently such a vendor will have a deep knowledge of the company about which the analyst is seeking information and may be able to point out elements of strength or weakness of which the company's officers themselves are not fully aware. One spectacular example of this was information furnished a security analyst about two companies each of which are internationally preeminent and both of which are among the Fortune U.S. 100 corporations. Both companies were facing the type of new product problem the solution of which is invariably found in time but delays are always highly costly. This vendor indicated that both companies had outstandingly capable engineers but that Company A would solve its problems in a considerably shorter time than Company B. This was because the A Company's personnel, if they gave an unconventional suggestion for solving the problem, were not the least afraid that their jobs or promotions would be jeopardized if in any particular instance their suggestion proved spectacularly wrong. In contrast, in B Company the engineers were sufficiently concerned that if they made a mistake they would be penalized that their tendency was largely to avoid the spotlight and keep their thoughts to themselves!

8. QUESTION: What have been your results over the last three years in reducing cycle time (total elapsed time from when the raw materials or components are delivered at the receiving dock to the time the product is ready for immediate shipment to the customer) and what further improvements do you see ahead?

COMMENTS: Reducing the cycle time, something on which enormous strides have been made in the last couple of years by alert companies, has several significant advantages. It obviously saves working capital. An unbelievably high expense for many companies is, unfortunately, theft. The shorter the period of time that materials are in the jurisdiction of the company the smaller the opportunity for such theft. Finally, it enables faster delivery to the customer without having to keep large inventories on hand. It is matters of this sort where big progress has been made by some companies in recent years that give important clues to which companies are progressing and which ones are not.

9. QUESTION: How many calls per quarter do you or people in your company's manufacturing arm make to customers to learn where your service to the customer can be improved?

COMMENTS: Because manufacturing is normally not on the front line of those calling on customers this question is not so much a check on manufacturing but on top management's attitude toward how concerned and ingenious they are in attempting to keep their customers pleased with their activities and even occasionally hitting the goodwill jackpot by performing beyond the customer's highest expectations!

10. QUESTION: So far as manufacturing is concerned, what capital expenditures are you planning over the next two or three years that should increase your profit margin? Are such expenditures primarily involved in more capacity to take care of an expanding market or are they of a nature that will reduce costs on what you are doing now?

COMMENTS: This question should always be accompanied by a statement by the questioner that he does not want to ask something that might be considered company confidential in that it would be of benefit should a competitor learn about it. If there are indications that a significant part of contemplated capital expenditures in the time ahead are of this nature it may be possible to get an idea of the growth that might lie ahead without in any sense jeopardizing such confidentiality by asking what percent of contemplated capital expenditures fall in such a class and what percent do not.

11. QUESTION: Trying to avoid the buzzwords "Japanese Quality Circles" what are you doing to stimulate *all* levels of manufacturing personnel to contribute both ideas and effort to improving the efficiency of the manufacturing operation?

COMMENTS: This is one of the most important, if not the most important of matters involved in judging the investment appeal of the manufacturing arm of a company. A discussion of this subject should lead to information on such matters as whether each relatively small production unit within the company is setting its own goals on results it expects to achieve in the period immediately ahead and whether the degree of attainment of these goals is publicized in relation to and in competition with other comparable units. Similarly, it should be ascertained whether meritorious results of individual small units are rewarded financially through some type of previously devised "profit sharing" plan, not rewarded at all, or an in between course is followed such as a company sponsored party for particularly successful units. A basic concept here that should be examined carefully is whether management believes that the majority of manufacturing personnel has sufficient confidence in overall management so that when a new and significantly different technique is proposed, all levels of manufacturing employees tend to follow enthusiastically, or whether there is a strong tendency toward foot-dragging and continuing doing things as they have previously been done.

12. QUESTION: What have you been doing to take advantage of leading edge technology to keep your own production facilities ahead of competition?

COMMENTS: Achieving lower costs, better quality or being able to satisfy customers by making variations of existing products in small lots at relatively slight extra cost all are ways that a company may benefit spectacularly from superior manufacturing skill. Technology can be a major tool in attaining such results. Because in only a few industries such as semiconductors is this a way of life, such needs are often not even recognized by either equipment vendors or their customers. Yet it is in industries where the pace of technological change is slower that the greatest competitive gains may be made by the company leading its industry in such matters. Therefore supplementing this general question, more specific questions such as the following might be asked when applicable: Ratio of pilot to production lines? Frequency of process patents that have been awarded or applied for? Are you working with vendors to develop improved equipment? Through vendors or other sources are you learning what competitors may be doing along these lines? Is top management aware of the competitive significance of these matters and are they giving recognition and rewards to those producing tangible benefits in these ways?

13. QUESTION: What other questions do you, Mr. Senior V.P. of Manufacturing feel should have been asked and were not? Are there any of these questions that you feel are essentially not truly important and should not have been asked in the first place?

Taylorism and Professional Education

JOHN E. GIBSON

Objective observers are becoming increasingly aware of the need to consider the manufacturing process as a whole rather than as an object for piecewise suboptimization. This holistic, or systems viewpoint must include manufacturers' relations with subcontractors and suppliers as well as customers. The manufacturing system certainly must include the interrelationship of the physical manufacturing environment, manufacturing management, and the worker. If manufacturing engineers and manufacturing operations managers are to contribute effectively to the redesign of the workplace, it seems obvious that their professional training must include a recognition of the new integrated manufacturing system reality and how to deal with it effectively. In this paper we consider how to adjust the training and professional value set given business managers and engineers to become consistent with this modern reality.

The American manufacturing environment is now in a rapid state of change. Yet, our business schools and engineering schools have not yet begun to provide the leadership that this restructuring of the American manufacturing environment demands. Some observers believe that American manufacturing managers have been late coming to the party, that they have been slow to recognize the advantages of Japanese and, to a lesser degree, European developments. I believe that a more fundamental question is why leaders in American manufacturing practice have received little or no help in their struggle from the professional schools where they must recruit the new generation of managers? As I see the situation, American business leaders are now well in advance of engineering and business schools in recognizing and practicing total quality principles, participative management, worker empowerment, and the like. If this perception is correct, why is it so? My answer will be that American professional school faculties have not abandoned Taylorism.

TAYLORISM

Frederick Winslow Taylor is high in the pantheon of American engineering heros (Copley, 1923). In his obsessive optimization of individual job shop operations, his disregard of the human side of enterprise, and his rigid separation of thinking from doing, Taylor is the paradigmatic manufacturing engineer. Taylor is important, not merely because he made revolutionary contributions to the manufacturing canon, but also because the general style he set became the universal paradigm for American engineering practice and for engineering education, and remains so even today.

I intend in this paper to focus on how the elements of Taylorism are applied in the workplace and in the engineering classroom and why this environment is no longer right for modern America. I hold that Taylorism continues to be a major obstacle in our path to manufacturing efficiency, and that it must be replaced as the central element of our engineering educational philosophy as well.

THE ESSENTIAL ELEMENTS OF TAYLORISM

What are the essential elements of "Tayloristic" engineering practice that currently inhibit technical progress? I suggest that the following seven are critical:

1. *Analytic* bottom-up approach, where analytic here is used in the classical sense of "breaking into component parts or elements."
2. The *absence of the goal-definition phase* in normal engineering design practice.
3. *Engineering practice in a vacuum, without regard to human factors.*
4. The *hierarchical, nonprofessional style* of current American engineering practice.
5. The fantasy of "*value-free design*."
6. The traditional Taylor practice of *separating thinking from doing*.
7. Strong emphasis on *individual reward for individual effort*.

Analogous to Taylor's procedure of breaking down the manufacturing process into elemental steps, the first step in the engineering design process is the careful division of the overall task into simple subelements and assigning these parts to individuals or teams for detailed design. This is so simple and obvious, and it works so well in certain practical design tasks and in engineering design education, that we may fail to understand the deeper implications of this step.

It is clear that the Tayloristic process works best if the boundaries of the subunits are sharp and well-defined and interconnections are clear and separable. When devices demand extraordinarily tight tolerances, however, tighter even than tens of thousandths of an inch, we cannot break such complex and precise devices into subunits and assign the design and production to different teams. Nor is this good practice where interconnections may outnumber the subunits, or where boundaries may be somewhat tenuous.

Furthermore, we cannot leave the manufacturing engineer out of the design process. High-speed, precision production requires that the designer and the manufacturing manager work together in a team with the materials specialist. In the colorful terminology of one of my manufacturing manager friends, we must "ask" the material what it wants to do and how it wants to behave. Then we must "ask" production machines how they want to make the part.

But these are only a few of the more obvious implications of the *analytic, "bottom-up"* Tayloristic approach to engineering. One other implication may be somewhat less obvious. *The classically trained engineering "bottom-upper" accepts the goals of a project as given.* Such engineering goals are embodied in the "specification sheet." How could it be otherwise? The classically trained engineer may ask. How can one design or manufacture something without a specification sheet or blueprint? This question may be perfectly logical when applied to a conventional, well-understood object but irrational when we face the unknown. By insisting on a well-developed and complete set of specifications before one can begin the design and production of a new and untried object, the engineer removes himself from the most exciting, creative step; helping to set the specifications in the first place. But this is exactly the way we currently teach engineers to think and to design.

In engineering education, the Tayloristic approach seems so obvious that it is universal. We begin with the simplest mechanisms and equations, then proceed step-by-step to more complex devices and mathematics, in a bottom-up manner. Thus, the budding engineer is taught without words to accept engineering reality as susceptible to decomposition into simpler subunits best handled in isolation, a hierarchical management approach with professors as "bosses" who "think" and students as "workers" who "do," and an absence of discussion of goals, except for questions that are meant to elicit what the boss wants.

An alternative and more reasonable approach to a new engineering problem, however, is the *top-down* approach, which moves from the general requirements to the specific. This is the essence of the systems method and it is a natural way in which to introduce engineering students, even freshmen, to the process of engineering design. But some engineering professors vehemently object to teaching engineering design in the first year or two of an engineering curriculum. They argue that without a complete understanding of the elements of design, that is conventions, protocols, and detailed analysis, engineering design cannot be done. Thus, they say, engineering design must be reserved to the final undergraduate year, as the capstone experience.

For bottom-uppers, design does not start with understanding the client's needs, or with the environment within which the object is to function, or with an examination of the way people will use the object, or with the plans for the retirement of the object. Instead, design begins for bottom-uppers after the specifications are set, and stops with physical manufacture. Some might argue to the contrary, however, that *the only truly professional element in the design process is interaction with the client to determine jointly the operating environment and specifications of the object to be designed.*

If engineering educators inculcate reverence for inviolate specifications, as we continue to do, we are also implying that goals are external to the design process and are to be set by someone else. This *absence of the goal-definition phase* is the second major distinguishing feature of conventional Tayloristic engineering practice that is crippling our national attempt to regain manufacturing leadership in world markets.

The third crippling attribute is *engineering practice in a mechanistic vacuum, without regard to human factors.* Human factors must enter into the design, production, use, and especially product retirement. Yet, none of these essential steps is considered currently in engineering education. Humans will use the objects we design and build, but we engineers easily divorce ourselves from responsibility to these human users if we can.

A fourth debilitating attribute of current American engineering practice is its hierarchical, nonprofessional attitude. Conventionally trained engineers *accept* that they do not have a say in setting specifications for the design object, or in how the product may be manufactured, or in providing for graceful retirement from service. They *accept* that they are not professionals with an overarching professional responsibility to society for their work. They *accept* the fact that they are employees and thus *should* be told what to do. And we engineering educators seem to agree. For the most part, we are not registered professional engineers, and we do not encourage our students to look upon themselves as professionals in training, with professional registration as the confirmation of professional status.

The fifth element in current American engineering practice that gives me concern may grow out of the dehumanizing attitude mentioned as number three above. It is *the fantasy that engineers are engaged in value-free design.* This can lead to the belief that designers and builders have no responsibility for the use to which our products are put.

One of the primary features of Taylorism is insistence on a *rigid separation of thinking from doing.* Taylor prohibited participation by production workers in the organization, planning, and direction of the manufacturing process. Taylor required his workers to do exactly as they were told to do and no more. This authoritarian stance is carried over into engineering education through its rigid exclusion of students from participation in the planning, organization, and direction of the education process. We all learn by example, and this is one of those debilitating attitudes engineers learn without being conscious of it.

Individual reward for individual effort in the workplace implies an emphasis on piecework, separate postproduction quality inspection, and a resistance to the team concept. For example, auto factory line foremen long waged war on any sort of worker interaction on the line. Even talking was forbidden in the early days, and this clash with the traditional American value of mutual support no doubt hastened unionization. In engineering education, this attribute causes us currently to focus excessively on individual student performance and active discouragement of student team formation. As a result engineering graduates have little or no experience in team building or cooperative effort. Thus, when they do run into the need for team effort, many engineers exhibit resistance, discomfort, and clumsiness at interpersonal professional relationships. Engineers feel the "need" to know who is the boss and for a strong management structure. The "leaderless group" leaves them distinctly uncomfortable (Gibson, 1981). Engineering faculty members often carry this individualism even further. I have been present at a number of faculty promotion and tenure committee meetings at which it was seriously proposed to discount publications according to the number of authors on the paper. Under this concept a two-author paper would find each author awarded half a publication, and so on.

Unconscious Taylorism in engineers is, I believe, responsible for the sabotage of many participative management programs.

WHEN DOES TAYLORISM WORK?

We know that Taylorism worked and worked well in the early part of this century. Can we be more specific about the economic or social conditions for which Taylorism is well suited? We should be able to examine the structure of American engineering curricula and ask "for what criteria are these programs optimum?" The idea is that the programs optimize for something, but not necessarily for desirable goals.

It is clear that for the most part, the faculties of accredited schools in business and engineering are hardworking and dedicated individuals, possessing high professional skills and a deep interest in transferring their knowledge to students. Students in business and engineering are unusually dedicated and hardworking and above average in learning ability. Furthermore, the programs in both sectors have traditionally been well supported and generally have adequate resources. Thus, how could it be that such programs might be seen as suboptimum or even ineffective and possibly deleterious to the future economic health of the nation?

As one examines business and engineering curricula, it seems clear that on a bit-by-bit basis they are well done. Each course by itself seems to have clear goals and effective procedures for producing course-by-course optimization. For this the accreditation process deserves praise. But this academic process is patterned after the old Tayloristic suboptimization of individual operations on a manufacturing line with no thought for overall production efficiency.

Suppose we examine the features of Taylorism we have discussed and attempt to describe the world for which they seem appropriate. The *bottom-up approach* should work well in optimizing a standard process. If the process is operating properly, *optimization of elements* will result in further efficiency. One need not discuss objectives *if the design objectives are universally accepted*, as they are in a traditional organization that produces a traditional product or delivers a traditional service.

Absence of concern for human factors is to be expected where working and living conditions are primitive, as in a frontier community. *Hierarchy appears when the worker is ignorant, untrained, uneducated, and the same is true for separating thinking from doing.* The *value-free fantasy* could develop if the values were so stable as to be taken for granted. Thus, it seems that Taylorism could work well if the products to be produced are conventional, the marketplace is stable and predictable, and workers are unskilled immigrants in whom the nation has made no educational investment. This does not describe the modern high-technology, rapidly changing world marketplace in which the United States must prosper. It does not describe a workplace that makes use of highly educated, socially advanced citizens.

WHAT SHOULD BE DONE?

If we agree that the level of Taylorism in engineering education should be reduced, the following series of steps could prove effective. First, stakeholders should be consulted to determine an appropriate set of goals, and quantitative performance objectives (metrics) should be established for professional education. Participation in the study should be solicited from industry and professional societies, the National Academy of Engineering, National Society of Professional Engineers, and professionals in practice.

The goals developed should not be limited to generalized platitudes. Rather the goals tree should be elaborated down to specific measurable, operational objectives. The elaboration needs to be carried down to the point at which it is possible to attach a quantitative performance measure to each subgoal. Measurement metrics must be developed and agreed upon, and then various solution options must be considered. This effort should not be aimed at dictating "the solution." To do so would be to slip back into Taylorism. Rather it should aim at producing IF-THEN scenarios. That is, *if* a certain option profile or plan is adopted, *then* the following outcome is likely. Furthermore, because education is a process, we must dedicate our efforts toward eliminating the idea of a static, fixed "truth" and methods of teaching it, and move toward emphasizing continuous improvement in providing a quality product, where quality is determined by our customers and not by ourselves.

Without attempting to prejudge the action plan to be produced, I will risk making certain specific suggestions for consideration. These suggestions seem naturally to fall into the following four general categories, for each of which I suggest more specific tactics.

- Empowerment of our professional students
- Encouragement of cooperative student work practices
- Participative management of the educational enterprise
- Development of a supportive professional accreditation process

Empowerment of Our Professional Students

Our students should be encouraged to take charge of their own learning. The following tactics suggest themselves:

- Encourage students to set and meet their own intermediate performance goals within each course.
- Provide optional homework packages instead of making all required.
- Let students take (computerized) examinations for self-diagnosis of knowledge gaps and provide suggested remedial practice material.
- Put in place a program of self-paced instruction.

Encouragement of Cooperative Student Work Practices

Engineering education has overemphasized the Tayloristic practice of exclusively individual work and individual rewards. This training has inhibited cooperation in the workplace by graduate engineers. Encouragement of cooperation here refers to cooperation not only among students but also among students and faculty members. Here are a few thought starters:

- Encourage teamwork on homework with one submittal and one grade for a group.
- Encourage advanced students to tutor beginning students for partial course credit.
- Establish courses within the engineering curriculum in which student teams solve industrial and community problems for real clients.
- Set up design juries of industrialists or national design competitions in required courses, thus turning faculty members into coaches and advisers for student teams under student leadership.

Participative Management of the Educational Enterprise

It is pure Taylorism to say to students that faculty members think, and students are the workers, thus they do (without thinking, unfortunately). How can we encourage engineering students to go into industry and help to manage participatively when they have been trained in an exclusively autocratic environment? Of course, overall course and program goals are set externally by the Accreditation Board for Engineering and Technology (ABET) and by employers. That is not the issue. But *how* to meet these performance goals is the proper subject of participative interaction. Hence, the following suggestions:

- Clearly establish necessary performance goals for each course.
- After working through necessary required basic material, allow students to participate in choosing from among optional blocks of material such as more theory, alternative techniques, and application areas.
- Appoint advanced students to course and program planning committees.

Development of a Supportive Professional Accreditation Process

ABET and its predecessor, the Engineers Council for Professional Development, have accomplished a great deal of good over the past 60 years, mostly with unpaid faculty volunteers. Now the professional accrediting process can be called upon to make further important contributions including the following leadership efforts.

- Eliminate micromanagement of individual course content in inspection process in favor of measurement of achievement in overall courses and programs.

- Involve ABET in developing coaching and training materials and programs for faculty in support of the conversion to participative management in engineering education.
- Shift ABET attention from "inspecting quality in at the end of the line," toward participative, cooperative empowerment of faculty (and students) in internalizing the unending quest for quality.

SUMMARY

It appears to me that Taylorism is alive and well in the minds of engineering faculty throughout the nation. Furthermore, it appears that the unresponsive, change-resisting attitude exhibited by many engineers in American manufacturing practice is in large measure due to this primitive and ineffective educational paradigm.

ACKNOWLEDGMENT

I benefited from participative deliberations and advice from several of my faculty colleagues, including K. P. White and my research associate R. Mathieu, and from about 40 graduate students who read earlier drafts of this paper and engaged in vigorous criticism and advice. I wish also to thank Dale Compton for his comments, which improved the focus of the effort. It seems fair to say that this work product was an effort of a team operating in the Hersey-Blanchard P-Mode.

The Integrated Enterprise

WILLIAM C. HANSON

The competitive and technological challenges that face manufacturers in the future demand that they operate in a broader, more holistic context. No company can succeed by itself. Those that try to do it alone may not survive.

The successful manufacturer will need to view itself from a new perspective. It will need to view itself in the broader context of a manufacturing enterprise and to understand that the factors that contribute to its manufacturing effort go far beyond the traditional production cycle. These factors encompass the entire range of activities that begin with market demand and end with customer satisfaction. Thus, the manufacturer must adopt a worldview in providing the utility that solves customers' problems. Developing this worldview begins by recognizing that even though all the company's internal organizations—sales, marketing, engineering, and manufacturing—operate interdependently, they have a common focus and are committed to delivering customer satisfaction. But it cannot stop there. This entity must be expanded to include outside organizations such as suppliers, consultants, research centers, competitors, and the customers themselves. Each of these groups has resources and knowledge vital to the ultimate solutions. They all must be integrated into a cohesive "enterprise" working toward shared objectives. It is this *Integrated Enterprise* that allows both the manufacturer and the customer to be successful.

The task of addressing all the internal and external elements of the enterprise as a cohesive whole, rather than as a set of discrete functions and organizations, raises some interesting issues. One issue is the interrelationships that organizations have and the changes necessary in these relationships to create a cohesive whole. This requires new thinking about how the various elements of an organization work together, organize, behave, relate, and measure; what they value; and how they are motivated. Timely solution of customer problems requires that a manufacturer have some collective unity and focus before a specific customer need is identified. Successfully addressing this interrelationship will allow us to develop the framework for an integrated organization. This unity will come by focusing on the operating excellence required by all customers and will be recognized through an operating vision that demonstrates that excellence. This vision encompasses the following elements:

- Products and processes that "never fail."
- Shortest cycle time in the industry.
- Competitiveness independent of volume.
- Leadership in defining industrywide manufacturing excellence.
- Leadership in the development of the best people.

This vision can be realized only in an environment that encourages a "learning process," the mechanism that draws knowledge from the disciplines critical for success. Finally, the elements of a manufacturing enterprise need to view themselves from the perspective of the task that must be accomplished, not the organization in which they are members.

For example, a key manufacturing metric is cycle time. In traditional manufacturing, cycle time refers to the period of time required to produce the product. That is the duration of the production cycle in creating a finished good. Within the context of the Integrated Enterprise, cycle time is redefined as beginning when the customer expresses a need and ending when that need is fulfilled. It includes time for problem identification, sales, order processing, supplier delivery, design, assembly, shipment, invoice, installation, and service. Each element in this process and its relationship to other elements must be considered in order to reduce cycle time. The "task" of reducing cycle times must be viewed in a broader context to identify all the variables that effect cycle time. Therefore, the definition of cycle time must extend beyond the traditional structures in order to encompass all the variables.

A second issue concerns developing the trust required to encourage a successful team orientation. What allows individuals to act as one, is having equal access to the information and knowledge that describe and justify the task. When people are excluded from information and knowledge, they should not be expected to act in unison. A supplier unfamiliar with marketing plans and product strategies cannot fully provide the resources and intelligence to help reduce time to market. Exclusion not only affects the potential efficient flow of the suppliers' material but also denies full access to their knowledge and expertise. In many cases this knowledge on improving time to market may be more valuable than the raw material suppliers provide. Although computer technology provides the means to facilitate the free flow of information, independent of geography or organization, there is still a reluctance to allow information across organizational barriers. Restricted information flows are a function of organizational behavior, not technology. The Integrated Enterprise requires that these barriers be eliminated.

The Integrated Enterprise creates the framework for the successful resolution of these issues. But the architecture is only the shell unless we add one more critical piece, namely, the free flow of information unimpeded by cultural and organizational barriers. These barriers are created when each element of the enterprise views itself as an end and not as a means to fulfilling the ends of the larger enterprise. Each discipline, seeking to optimize its own operation creates its own culture, somewhat independent of other organizations in the company. In this regard the organizational structure itself can become the barrier. What must change, therefore, is how the activity is viewed within the organization, which must accommodate that view in as efficient an organizational structure as possible.

The challenge is to adopt a structure that is organized around a stream of activities that transform knowledge and material into customer solutions. This implies a radical change from the traditional functional organization. The successful manufacturer of the 1990s will not have separate organizations devoted to discrete functions such as manufacturing, engineering, or marketing. Rather it will be structured and viewed in terms that relate to the customer. Customers do not buy manufacturing, engineering, or sales; they buy solutions that fill needs. The successful manufacturer will focus the organization on customer needs, not on the functional capabilities of the organization. In this way the entire enterprise is optimized around meeting the customers' needs, using the skills of each discipline, focusing on the real task, and ultimately solving the real problems.

Focus on the customer versus the function does not reduce the need for excellence in the traditional disciplines. Companies must continue to require such excellence, while developing the strengths, skills, and knowledge of each individual. What is different here is that each set of skills must be treated as an embedded discipline of excellence rather than an organizational structure. The problem is that companies continually think of these areas of skills as functions. Functions unfortunately connote an organizational structure. But if these skills could be viewed instead as disciplines, as resources to accomplish tasks, companies could then free themselves from the traditional barriers that surround organizational structures.

An additional benefit in focusing on the real task of providing the optimum customer solution is that it makes clear where resource gaps exist and which skills and information are lacking. What is implied here is that the walls of the company become limitations to success. Outside suppliers, customers, academia, consultants, and even competitors can add significant value and knowledge in accomplishing the task. These must be viewed as critical additional resources that provide valuable goods, services, or information that help define and deliver greater utility in supplying the best solutions.

Implementation of a structure and articulation of a set of values are the foundation of the Integrated Enterprise. A successful implementation exhibits five principles of leadership in the management of people and technology. People leadership refers to cultures and values. Technology leadership addresses the integrated information systems that allow the electronic interchange of data and the sharing of information, knowledge, experience, and values.

The first principle asserts that when people understand the vision, or larger task, of an enterprise and are given the right information, the resources, and the responsibility, they will "do the right thing." Doing the right thing depends on the appropriate frame of reference and a clear understanding of the task and its scope. The Integrated Enterprise, working with shared objectives, ensures that people are doing the right thing in the proper frame of reference. In this context it is critical that the members of the enterprise freely share a common understanding of the task. For example, if the problem is shorter cycle times, then a company must share that information with its suppliers. Product specifications, volumes, and scheduling within the context of broader product strategies that include inventory strategies, quality control, customer lead times, and distribution plans are examples of the information that should be freely shared. When suppliers can make proper decisions that contribute to reducing cycle times, they become part of the team that successfully completes the task.

The second principle addresses empowerment of the individual. Empowered people—and with good leadership, empowered groups—will have not only the ability but also the desire to participate in the decision process. This level of involvement will enable and encourage the individual to make decisions rather than adopt a passive or reactive attitude, waiting to be told what to do.

The existence of a comprehensive and effective communications network is the third principle of the Integrated Enterprise. This network must distribute knowledge and information widely, embracing the openness and trust that allow the individual to feel empowered to affect the "real" problems. But the network alone is not enough. The democratization and dissemination of information throughout the network in all directions irrespective of organizational position becomes a critical fourth principle that ensures that the Integrated Enterprise is truly integrated.

The results of the first four principles imply the fifth—distributed decision making. Information freely shared with empowered people who are motivated to make decisions will naturally distribute the decision-making process throughout the entire organization.

The principles of the Integrated Enterprise will compel companies to reconsider their organizational strategy. By capitalizing on the democratization of information and by distributing the decision-making process, a company ensures that decisions are made where the work occurs. Note that much of the work will be outside the traditional boundaries of the manufacturer. In this context relationships become peer-to-peer, not hierarchical. People share information to accomplish tasks. The organizational structure and the behavior that results from that structure become the key element in the effectiveness of the total enterprise. Interestingly enough, the most radical change must begin inside the four walls of the manufacturer itself. Company structures must dramatically change to capitalize fully on the opportunities provided by the Integrated Enterprise. Much of our segmented and restrictive thinking begins at home.

Consider the typical hierarchical company organization chart. The standard configuration would define relationships between supervisor and subordinate, placing operations at the bottom. The word *operations* here refers to all direct value-added operations, such as selling, designing, building, and servicing. Since a company's most valuable knowledge base is its people, and the majority of people are in operations, the operating groups logically should be self-managed and empowered. Unfortunately, companies have historically layered people (overhead) on top of operations to tell operations what to do and to mediate behavior between operations. Now if we visualize turning that organization 90 degrees, a new image, or perspective, is created. Each of the operating units is now seen servicing the needs of the larger enterprise and not the overhead layers and the functional segmentations they represent (see Figure 1).

Consider the dramatic change in the dynamics of relationships. The peer-to-peer relationship is of the kind that exists between a customer and a supplier. The dependencies are management by dialogue and negotiation. Knowledge and understanding are both intrinsically and institutionally far more important than rank. For example, resolution of problems in the delivery of a specific product or service solution is knowledge based, not position based.

Does this mean that the role of management is eliminated? On the contrary, managers must develop a new set of skills. Traditionally, the emphasis was primarily authority and decision making at a functional level.

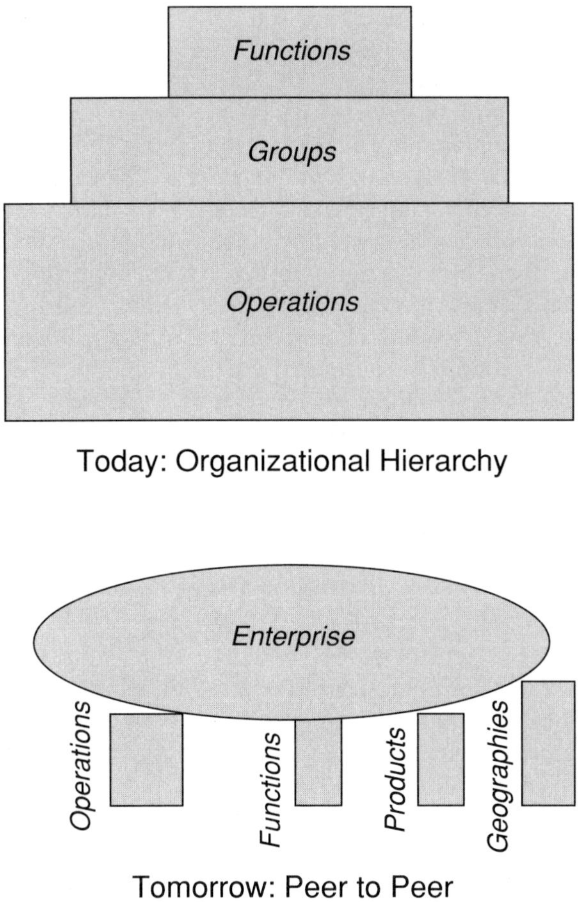

FIGURE 1 The organizational hierarchy of today transformed into the peer-to-peer structure of an Integrated Enterprise.

Now, leadership, task definition, and resource development become the critical strengths of the new manager. Leadership should not be confused with decision making. Leadership plays the critical role in ensuring that the enterprise understands the right tasks. Good leadership must clearly define the tasks, develop the resources, and create the positive environment that allows those resources to fulfill the mission.

It is important to reiterate that resources extend beyond the walls of the company. Leadership also requires that noncompany resources such as suppliers and customers are properly integrated as part of the solution. These traditional external resources are more easily integrated because of their place in the traditional value chain. But as the concept of value added is expanded to include knowledge and learning, we will quickly see the need to embrace a wider resource base, including government, academia, industry resources, and even competitors.

World-class manufacturers will be recognized by the leadership they provide in attacking and resolving complex customer problems. The effort needed to understand, define, and resolve these problems in an increasingly complex world intensifies the need for managers to acquire new skills for the 1990s. Complex problems will not be solved by simple solutions. Most often these problems present themselves as representing irreconcilable positions. For example, customers have always wanted high quality and low cost. But the historic conventional wisdom resolved the "dilemma" with an "acceptable quality level." One simply determined a desired quality level and invested the appropriate cost. Another example is the trade-off between short lead times and high inventories. The dilemma, how to got both high quality and low cost, how to have short lead times and low inventories, was ignored. This is no longer true. High quality and low cost, short lead times and low inventories are essential components of success in modern industrial life. Total quality management and just-in-time methods are recognized standards of excellence that result from resolving these historic dilemmas.

The successful manager of the 1990s will have the skills to define complex dilemmas and resolve them, not ignore them. This management skill, which can be defined as "dilemma management," is a critical component of the Integrated Enterprise. The characteristics of the dilemma manager include the ability to tolerate ambiguity, to manage and, indeed, thrive on the tension that is caused by apparently conflicting demands. The apparent conflict will be valued as a stimulator for change. It will encourage new levels of creativity, and the resolution of one dilemma will create new dilemmas to solve. Once you get to one level of performance, it will be time to move to the next level. Patience and courage will be of premium importance. The successful manager will value the dilemmas and have the vision to stimulate them, not eliminate them by trying to make a trade-off between good opposing views. The emphasis is on continuous improvement. Those manufacturers whose managers can solve industry dilemmas first will have a competitive advantage.

The world-class manufacturer of the 1990s will be an integrated enterprise. It will capitalize on a wider audience of skills, beyond the department, beyond the walls of the company. It will engage in strategic collaborations that extend beyond the organization and will capitalize on increased transfers of technology and knowledge. It will encourage the free flow of information throughout the enterprise and will empower large numbers of people to work cooperatively in a peer-to-peer relationship that encourages and motivates dynamic distributed decision making. It will embrace "dilemma management" and, by identifying and resolving dilemmas, rise to higher and higher levels of performance. It will value and demand change that will yield continuous improvement, setting ever-higher standards of performance. It will be recognized as defining the standards of excellence to which others aspire.

Time as a Primary System Metric

DAN C. KRUPKA

Time—the interval from the start of manufacturing activity to its completion—is the single most useful and powerful metric that any firm can employ to measure its manufacturing operation. This paper argues that time is a more useful and universal metric than cost and quality because it can be used to drive improvements in both.[1] If we were offered a second choice of metric, we would add the variance of that interval. The traditional view is that cost, quality, and time are the important elements for assessing manufacturing performance. Here it is argued that properly managing time will ensure that the other two metrics fall in line.

To begin, it is necessary to recognize that manufacturing operations—the activities that take place within the walls of a factory—can no longer be treated as the system to be optimized. Instead, in considering a manufacturer, we must think of several systems, of which manufacturing is but one (see Figure 1). Customers' orders for products are conveyed by an ordering system to the manufacturing system, whose output then flows through a distribution system to the customer. Rapid and flexible response requires that materials and parts flow quickly into manufacturing; that requires a

[1]In making this suggestion, I am echoing George Stalk, Jr., and Thomas M. Hout who advance this argument in their book, *Competing Against Time*. Many of the ideas presented in their book become even more convincing when considered in light of the concepts discussed in the section on lessons from queuing models.

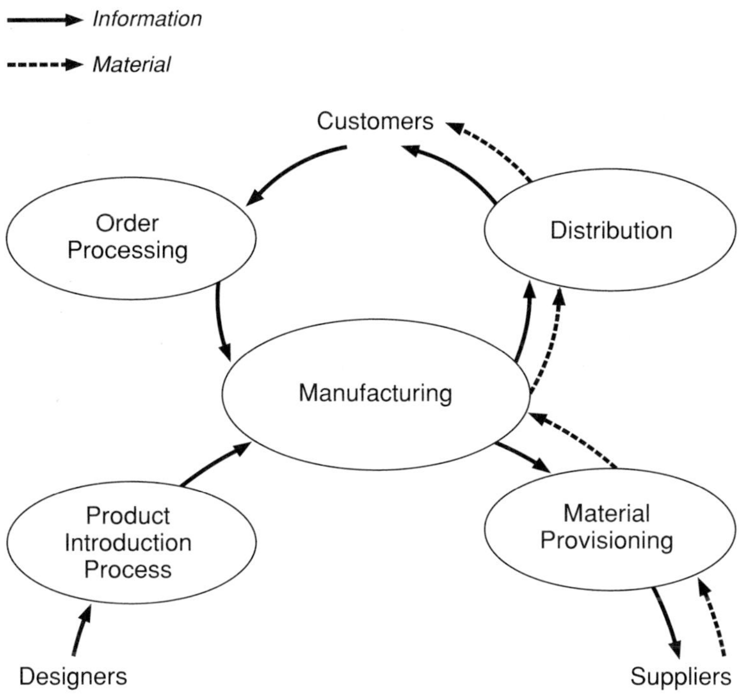

FIGURE 1 Targets for interval reduction. Several systems are needed to define a manufacturer besides the activities that take place within the factory walls.

short and predictable interval for the material provisioning system. In addition to ensuring high performance for these systems, a successful firm must be capable of rapidly translating its designs into manufactured products. Hence, we need a well-crafted and rapid product introduction system.

All of these systems bear great similarity to one another. Each consists of a sequence of processes and prescriptions for defining how entities—materials, products, or information—are to flow from one process to another. Currently, manufacturing processes on the factory floor are better understood because they have a long history of analysis and design. As many are finding, however, the other systems considered here can be analyzed and improved by employing methods analogous to those applied on the factory floor.

To manage these systems as part of an integrated whole requires a metric that ties them together operationally. Time is that metric. In contrast, there is no common definition of quality throughout the system. On the factory floor, we speak of defect levels, yields, or rework rates. Although analogous terms may be applied to processes in the other systems, it is conceptually difficult to compare, for example, the severity of a defect in a customer's order with a defect on a silicon wafer. Not so with time.

What about cost? The difficulty with using cost to measure *the performance of the system* is that it is a lagging metric. Books are closed monthly, not daily; the calculation of costs always requires arbitrary allocations of expenses. Costs, as traditionally calculated, are volume dependent. Reducing time (and its variance, as will be discussed later) is always a benefit; cutting costs is not. Consider for example, the consequences of reduced staffing in a pilot production facility, thereby unwittingly creating a bottleneck in the product introduction process. A significant delay here may result in dramatically reduced profits over the life of the product (Reinertsen, 1983). Thus, it is difficult to make meaningful comparisons between the *capabilities* of manufacturing systems using cost as a metric.

The foregoing arguments may sound academic. They are not. Competing against time is fast becoming today's business strategy. According to Stalk and Hout (1990), compressing time has been observed to lead to the following sequence of changes: productivity rises; prices can be increased as responsiveness to customer orders is improved; risks are lowered because reliance on volatile forecasts is reduced; and market share increases as a result of improved responsiveness. In light of these very practical arguments, the measurement of time throughout an enterprise is critically important.

TIME AS A DIAGNOSTIC TOOL AND A DRIVER OF QUALITY AND COST

Not only is time an excellent metric for the overall system (Manufacturing with a big M) and its subsystems (such as factory floor operations); it can be used to guide activities to improve performance. Moreover, reducing the interval, be it in product introduction or in manufacturing, will improve quality and cost. To use time as a diagnostic tool to improve a system, it is invariably profitable to begin by creating a diagram of the processes that make up that system. For manufacturing processes that have been designed by engineers, such diagrams are generally available. Such is not the case for product introduction or other nonmanufacturing systems. And yet these systems are often more complex than those encountered on the factory floor. Analysis of such diagrams often reveals the presence of steps that add no value or that consist of re-creating—at the risk of introducing errors—information created elsewhere. Eliminating these steps will shorten the system's interval, reduce costs, and often improve quality by reducing opportunities for the introduction of errors.

The diagram will also reveal opportunities for concurrent execution of activities. Introducing parallelism into a system that had consisted of serial process steps often carries benefits that extend well beyond the time saved.

An excellent example is product introduction as practiced by Japanese automobile manufacturers. Their cycles are considerably shorter than those of their U.S.-based and European competitors (Clark and Fujimoto, 1989a) because they have replaced a phased approach in which the activities of manufacturing engineers follow those of the product designers by an overlapping approach: Cross-functional teams are established early in the process, allowing preliminary information created by the designers to be assessed by manufacturing engineers. As a result, the overall product introduction interval is shortened, leading to lower development costs and a more manufacturable product possessing higher quality. A further payoff is that firms with the ability to introduce products rapidly can be much more responsive to market trends.

LESSONS FROM SIMPLE QUEUEING MODELS

Despite the inability of many firms to create systems characterized by the virtual elimination of non-value-adding steps and by the introduction of much concurrency, the prescriptions outlined in the previous section are straightforward. Even if a firm had made all the suggested improvements, however, there would remain opportunities for reducing its interval. Insights from elementary queueing theory show how relentless concentration on the time metric leads to additional improvements in performance.

Figure 2 shows the average throughput time of a single-server queue (such as that associated with a single machine on the factory floor) as a function of the system's capacity, measured as the ratio of the arrival rate of entities to the rate at which the resource (or machine) is able to perform its function (Whitt, 1983). (Purists would prefer more careful definitions. My intent, however, is to sacrifice technical rigor for simplicity of exposition.) The important point is that, for a given capacity utilization, the throughput time depends on the degree of variability in the system. In fact, if all variability were removed, the throughput time would be equal to the time required to perform the designated task—the so-called service time—until the arrival rate exceeded the service rate. At that point, a queue would form and grow without limit.

It is important to note that the ordinate on Figure 2 is the *average* throughput time. The throughput time fluctuates and, as one would expect, the lower the variability in the system, the lower the variance of the throughput time. Consequently, to assess the performance of a system, we should measure not only its throughput time but also its variance.[2] How does the variability arise? We distinguish two classes of sources: those affecting

[2]This point is also made by Hopp et al. (1990). Their paper suggests incorrectly (p. 80), however, that queue time can be directly addressed and is unaffected by setup time.

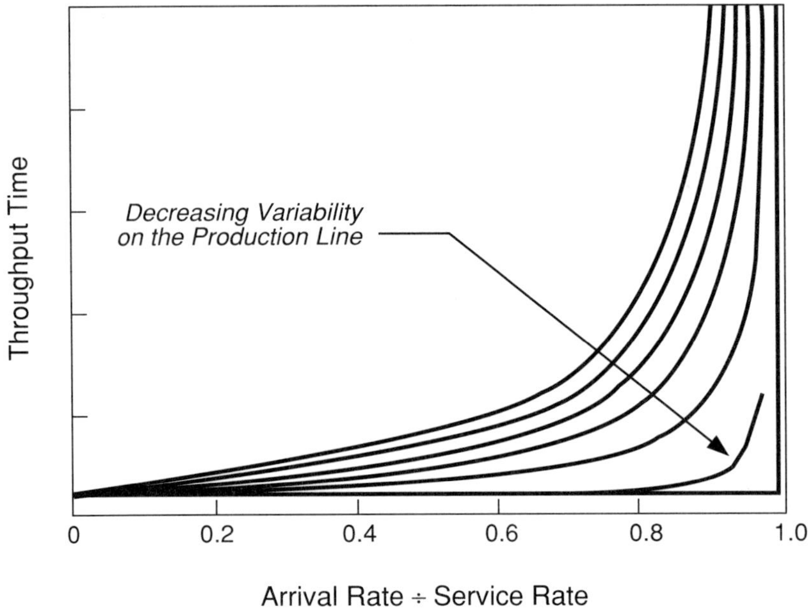

FIGURE 2 General results from queueing theory. Here the average throughput time for a single server queue illustrates the effect of variability in the system.

the arrival rate and those affecting the service rate. On the factory floor the arrival rate at a given stage may fluctuate for any number of reasons—for example, arrivals from upstream stages that are not in phase, problems encountered at the preceding stage, equipment failure, and random batching. The service rate is affected by problems experienced with the equipment, variable setup times, lack of clear instructions, or lack of operator skills.

It is important to observe that at levels of capacity utilization that make economic sense, say in the neighborhood of 0.8, small decreases in the service rate (which effectively shift the system to a higher level of capacity utilization) lead to a large increase in throughput time. In addition, an increase in variability in either the arrival or the service rate, leads to a large increase in throughput time.

Although the curves shown in Figure 2 are calculated for a single-server queue, analogous results are obtained for a network of queues, such as a manufacturing facility or the ordering process for a complex product. In general, therefore, the prescriptions for reducing throughput time (or manufacturing interval) are the same: reduce variability in the system and strive to increase the service rate.

How do these prescriptions translate into quality and cost improvements? The answers are straightforward, at least in principle. Reducing the interval by reducing variability requires that the sources of this variability be systematically eliminated. This requires reducing rework rates and machine downtimes, smoothing the flow in a manufacturing line by reducing batch sizes and by appropriate sequencing, devising approaches to minimizing shortages of materials, improving operator skills, ensuring that bills of material are accurate, or improving the accuracy of storeroom on-hand balances. All these, and many others, are activities that are addressed in a typical manufacturing environment and which are measured with a variety of metrics. By thinking of these activities as being aimed at reducing variability, we see how time and its variance, as high-level metrics, drive improvements in quality and cost.

The foregoing discussion glosses over an important concept—translating the high-level metric of time into the performance of a specific activity, be it on the shop floor or at a stage in the product introduction process. For example, management cannot demand that operators reduce the manufacturing interval on their line. High-level, time-based goals must be translated into the activities that will support those goals. And the responsibility for performing that translation lies with management.

Although it should be obvious from the foregoing discussion that reducing variability reduces costs, it is worth considering an example that dramatically illustrates the point. A manufacturing manager wished to confirm the need for an extremely expensive machine. The factory already possessed one such machine, which was fed by the outputs of two different lines, but its queue time had become unbearably long. Upon investigation, it was discovered that the material handler responsible for moving the product from the two feeder lines to the expensive machine wished to minimize the number of trips that he made. He preferred to transfer the product in large batches rather than moving it as soon as it arrived at the end of the line. Since the expensive machine was already highly utilized, the materials handler's strategy had a devastating effect on the line's performance. By changing the material handler's schedule, the need to spend an additional $1 million was avoided.

Earlier, with regard to the potentially unfortunate consequences of cost reduction in pilot operations, it was suggested that reducing time is always recommended. Such a prescription requires careful interpretation because mindless pursuit of shortened intervals by the addition of excess capacity could lead to higher costs. Until the importance of speed is more widely acknowledged, however, it is excessive zeal in cost-cutting and the neglect of the more fundamental parameter, time, that poses the greater danger to the majority of firms.

The discussion in this and the previous section demonstrates that time, as a primary metric, is the ultimate detector of inefficiencies in an overall system. It quickly uncovers flawed performance. For example, if the time that elapses between placement of a customer's order and receipt of payment is much longer than the manufacturing interval, the overall system is clearly inefficient. Even on the factory floor, high first-pass yields—a favorite measure of overall performance—may lull us into complacency. A far better measure is the manufacturing interval or the ratio of value-added process time to the manufacturing interval. The latter allows one to determine how close the process is to reaching the theoretically minimum time.

APPLYING THE LESSONS LEARNED IN MANUFACTURING

In the opening section we observed that the nonmanufacturing systems such as product introduction bear an important similarity to the shop floor. Since more sophisticated approaches such as queueing analysis or simulation have been applied primarily in the manufacturing environment, are there less obvious concepts that can be ported from the factory floor? The answer, of course, is yes.

Perhaps the most important lesson is that these nonmanufacturing environments should be conceived, at least qualitatively, as systems through which entities should flow rapidly. One area in which firms often economize is pilot production. In light of the arguments presented here, that may be a false economy.

We may also draw a few additional lessons. Just as flow through a manufacturing line is facilitated by small lot sizes, so speed in product development can be achieved by aiming for small increments in product capability and introducing these often (Gomory and Schmitt, 1988). Whether or not they draw the analogy with lot sizing, Japanese manufacturers have skillfully applied this concept, particularly in consumer products. On the factory floor, performance is enhanced by organizing manufacturing into focused factories. In product development, the analogous response is the establishment of cross-functional teams. In an ideal line, product, once started, flows automatically and requires no scheduling. To achieve rapid product introduction, the design should move ahead once the specifications are frozen; there should be no reopening of specifications. Also, just as short manufacturing intervals are facilitated by total quality control, so rapid product introduction depends on the quality of underlying processes.

Finally, just as engineers are essential to just-in-time and total quality improvements on the shop floor, so are they essential to engineering the other systems. In fact, we are now seeing the birth of a new branch of engineering—an offshoot of industrial engineering—that we might call industrial operations engineering.

Communication Barriers to Effective Manufacturing

JAMES F. LARDNER

Empirical evidence suggests that important communication barriers exist between many of the functional groups in American manufacturing companies. There is also evidence of cultural and environmental barriers to timely, effective decision making by these organizations. These barriers prevent companies from responding rapidly and effectively to changes in market requirements and customer preferences. They also cause serious quality problems, raise product costs, and inhibit the ability of a company to compete effectively. This paper examines some of the causes of the barriers and discusses what needs to be done to overcome them.

One of the most publicized barriers is that between product design and production. In a process that has been termed "throwing-it-over-the-wall," design engineers concentrate on developing product functions and features with little or no consideration for how the product is to be made. Only when the design and development of the product are complete and the results are "thrown over the wall" can the production organization determine whether the product can be made at an acceptable level of quality, cost, and capital investment. The results can usually be characterized as follows:

1. A long time elapses between product concept and production.
2. The product exceeds cost goals.
3. Quality goals are not achieved during early production, and be-

cause production and purchasing people were not involved from the start, may never be achieved or, if so, not economically.

4. Excessive investment is required to produce the product at the quality and volume levels planned.
5. Early product life is marked by a large number of design changes—about half for reasons of quality and half for reasons of producibility.
6. No one accepts responsibility for failure to meet program objectives.
7. It is impossible to determine where responsibility for unsatisfactory results lies.

Though the relationship between design and production has received the most attention, other barriers are created by the "over-the-wall" process. One of these relates to material sourcing and procurement, a process commonly called purchasing. With few exceptions, final material sourcing and procurement decisions can be made only when the completed product design is thrown over the wall by the design group. The results here are similar to those caused by the barriers between design and production:

1. Longer lead times than expected and planned are necessary to get delivery of purchased material.
2. Product quality objectives are not met.
3. Product cost targets are not met.
4. A large number of changes in material sourcing occur early in the life of the product.
5. Unnecessarily large numbers of nonstandard materials, parts, and assemblies are incorporated into the product, inflating product cost and increasing the cost of product support.
6. Relationships with suppliers are adversarial rather than cooperative, and it is difficult or impossible to take advantage of a supplier's experience or technical capabilities during product design and development.

A third, less publicized, over-the-wall problem involves aftermarket product support. Service engineers are rarely involved early in product design and development. Mostly, they participate only during the final phase of product test and evaluation when they are asked to prepare product service information and recommended service parts lists for product support. Such late entry into the design and development process makes it impossible to capture critical input from service engineers based on their years of experience with past products and a great deal of knowledge about customer concerns, priorities, and use patterns. This failure results in products that are inconvenient and time consuming to service and expensive to repair.

The growth of manufacturing companies from small shops with limited product lines employing highly skilled workers to large multiproduct enterprises with substantial concentrations of capital and labor brought a high degree of specialization and division of labor in the work force. This change created more and more narrowly focused functional groups and resulted in a significant increase in organizational complexity and in difficulty of managing the manufacturing process.

To coordinate and direct the activities of the rapidly proliferating groups of specialists, additional layers of managers were added. As this was done, management became increasingly hierarchical and autocratic. Decisions were made at the top and transmitted down to the shop floor through the functional groups because that was how management was organized. Results were reported back to the top by the same route. It became increasingly difficult to maintain effective horizontal communication and coordination at any level in the organization except at the very top.

Over time, functional groups developed their own objectives and goals and evolved separate value systems. Each group dedicated itself to the optimization of its own function with little or no regard for, or understanding of, its effect on the performance of the manufacturing whole. Performance standards and reward systems varied from group to group according to group objectives and focus.

In this environment, every interface between functions became a potential barrier to effective communication and coordination. Clearly, this is a large problem, considering the number of functions in a typical manufacturing organization, the most important of which are as follows:

- Marketing and sales
- Design and development engineering
- Manufacturing (production) engineering
- Maintenance and plant engineering
- Sourcing and procurement (purchasing)
- Shop floor management
- Production work force
- Product service (aftermarket support)
- Accounting
- Human resources

The erosion of what once had been a common manufacturing language also created barriers to communication and coordination. That erosion was caused by the developing cultural differences among functional specialties. Over time, managers of each group began to edit and selectively interpret orders coming down through the organization. It is not uncommon today to find that each functional group describes and defines the product and the processes used to make it very differently. Although this practice does not interfere seriously with vertical communication within each function, it reinforces the barriers to horizontal communication and coordination between the various groups at every level in manufacturing.

The disappearance of a common language among the many groups in manufacturing highlighted a previously unappreciated problem in data and information management. This is the task of translating data and information from the root sources into the format and language needed by functional groups without losing the precise intent and meaning of the original. It is apparent that there is a serious lack of discipline in the data and information management systems used by most manufacturing companies and that this lack of discipline perpetuates barriers.

There probably is little hope of reducing the number of languages used by the variety of functions in a manufacturing organization. Therefore, providing a basis for common understanding of what the root data and information mean is essential to establishing a successful manufacturing process. To provide this basis, it is critical to be able to retain the original meaning and intent of the root or seminal data and information when derivative data and information are being created by functional groups for their own use. It is evident in practice that the act of translation is the problem and that the impact of imprecise or inexact translation on the entire manufacturing process warrants much more attention than it has been given to date.

Further barriers result from careless or inadequate definitions of key terms used in manufacturing. Quality is one example. Both practical experience and the technical literature argue that quality can have a wide range of distinct aspects and that their number, nature, and relative importance vary from product to product. Thus, a scoop shovel and a 757 airplane may have some fundamental quality aspects in common, but between the products, the more detailed aspects of quality clearly differ in number, importance, and kind.

A similar problem arises when defining costs. Although aggregated costs are similar whether the product is a shovel or an airplane, the way costs are defined and measured and distributed among the various functions of each manufacturing organization varies greatly. Because specific costs are seldom properly associated with the activities that gave rise to those costs, functional groups frequently establish goals and objectives that are at cross purposes with the primary objective of the whole manufacturing organization, namely, to produce at the lowest possible cost a product that meets all the market requirements.

If the cycle of design, development, production, marketing, and product support is to be shortened and manufacturing made more efficient, all the various activities that make up manufacturing will have to be reintegrated. This implies the need to perform many of these activities concurrently rather than sequentially as they are performed today. That will demand a highly interactive, intensively iterative environment, particularly during the development of a new product. Unless there is a data and information management system that is accessible to all participants and ensures that every participant has comparable data and information in regard to content and currency, this degree of integration will be impossible to achieve. Existing barriers will become even more damaging to the performance of the organization.

The general lack of satisfactory data and information management systems has encouraged the fractionalization of manufacturing. A manufacturing organization must react continually to changes in product requirements, product mix, product design, process design, material specifications, competitive pressures, and on and on with only brief periods of relative stability. Because of inadequate overall data and information management systems, functional groups have developed local systems in an attempt to maintain control over their own limited areas of responsibility. Since objectives and values vary from group to group, and there is little or no understanding of how the actions of one group will affect all the other groups, responses to changes in the manufacturing environment vary greatly. It is almost by accident that group actions are directed toward optimization of the whole manufacturing effort.

Manufacturing deals with continuing change, and change creates ambiguity. This is particularly acute during the process of design, development, and production. At the early stages of any program, data are scarce and subject to considerable subsequent alteration and revision. Things such as cost estimates, results of product test and evaluation, process analysis, and competitive activities continually affect design and production decisions. As the design and development process continues, however, more and better data become available and are less subject to alteration or revision. Until this begins to happen though, the absence of adequate amounts of good data inhibits decision making by project participants.

The lack of good data is a serious problem and contributes to the creation of barriers to effective project management. The problem could be minimized if there were a means of synthesizing data early in the cycle and then substituting hard data as the project progresses. The closer the synthesized data are to the ultimate "good data," the fewer barriers there would be. This might be possible using past experience and research-based approximations. Unfortunately, there are almost no good tools to help people do this. *Even though many of the decisions to be made are similar or identical to decisions made in the past, no history of those past decisions is kept, nor are the results of those decisions analyzed and evaluated.* Thus, nearly every current decision is made based on whatever may be remembered from the past and on whatever perceived results were believed to have occurred.

The uncertainty this creates is a cause of barriers to timely decision making and to achieving the best possible solutions. Capturing what has gone before and modeling the results, together with a broader ability to simulate results of certain proposed actions would not only speed up the decision-making process but produce better decisions as well.

Much decision making in manufacturing includes a high degree of uncertainty about the precise nature of the problems being addressed and about the likely result of any proposed solution. This is because of the essential nature of manufacturing itself. It is a monolithic entity, infinitely complex in all of its details but with each part so interconnected and interdependent that no part can be changed without also affecting every other part. Except for experience, there is no good way to predict the degree of change throughout the system as a consequence of a change in one of its parts. This constant uncertainty about the system response among decision makers is an important impediment to good, timely decision making.

There is a real need for a model or models to aid in determining the possible effects of various decision choices on the manufacturing whole. Development of a more accurate, better disciplined process than depending, as we do now, on the experience of a few key people and what they remember from past product programs would help greatly to improve communication and understanding among the various functional groups. It would also reduce the amount of iteration required and cut the time to arrive at acceptable solutions. With experience, it might be possible to model the interaction of increasing numbers of the various segments of manufacturing. If the effort led only to a deeper understanding of the dynamic interaction of more activities within the manufacturing whole, it would be worth doing. Time barriers to decision making are a major handicap when trying to react quickly to the market and to perceived competitive threats. A greater understanding of the possible results of any given decision could speed decision making and improve the possibility that the decision would provide a nonmalignant answer to a problem.

As the rate of change and the complexity in manufacturing increases, there is a growing need for multidisciplinary approaches to problem solving. In a product development program that extends over a comparatively long period, the number of people involved and the kinds and numbers of different disciplines and skills required vary considerably from time to time. At each point in the process, some players enter and some drop out. Others find that their status in the project team or the relative importance of their contribution changes as the project moves from stage to stage.

Unfortunately a large component of the education required for manufacturing management is experience based. Given that the traditional environment in manufacturing for the past 100 years has been one of increasing specialization and narrower focus, there has been no effective model to train manufacturing people to view and understand the integrated manufacturing whole. There is a need today for individuals who have broad knowledge and appreciation of the total spectrum of specialties beyond their own and how those specialties fit into the manufacturing system. Few institutions that train people who enter manufacturing industries provide a comprehensive understanding of manufacturing as an integrated system. This lack of an integrated systems approach to manufacturing is a continuing barrier to better management of the process.

In summary, a considerable number of barriers to effective manufacturing are related to the inadequacy of the current data and information management system and the present data and information structures themselves. (Data and information management, in this case, means the creation, storage, transmission, transformation, derivation, and manipulation and interpretation of data and information.) Since data and information are what drive and control the manufacturing process, it is critical that the management of data and information be accorded a high priority.

There is also a serious lack of understanding by those who work in the manufacturing system of how the system works. If there really is a generic data and information structure on which the manufacturing process depends, it may be possible to guide and manage the changes in individual and functional contributions and to establish common goals and understanding by using data and information management as an tool for integration.

In a sense, data and information may be likened to electric power. It is generated and distributed throughout a system. It drives various devices (functions) that perform work within the system. It can be transformed into various forms as required to perform the work, and it must be available instantaneously throughout the system in greater or lesser intensity and quantity according to the need. As it is with electricity and power-consuming devices, without data and information to drive them, none of the multiple activities of a manufacturing system could function properly and many could not function at all. Electric power by itself has molded the physical aspect and characteristics of the manufacturing industries. More effective management of data and information could mold the intellectual component of manufacturing to create more efficient, competitive companies.

Are There "Laws" of Manufacturing?

JOHN D. C. LITTLE

If we are to have a meaningful discipline of manufacturing, it is argued, then we should have intellectual foundations to include what might be called "laws of manufacturing." What are the prospects for identifying and creating such?

It may be useful to distinguish between three types of potential "laws": mathematical tautologies, physical laws and their analogs, and empirical models. Then we can ask separately whether we are likely to develop further along each line.

TAUTOLOGIES VERSUS LAWS

$L = \lambda W$ ("Little's Law") is a *mathematical tautology* with useful mappings onto the real world. $L = \lambda W$ relates the average number of items present in a queuing system to the average waiting time per item. Specifically, suppose we have a queuing system in steady state and let

L = the average number of items present in the system,
λ = the average arrival rate, items per unit time, and
W = the average time spent by an item in the system, then, under remarkably general conditions,

$$L = \lambda W. \tag{1}$$

This formula turns out to be particularly useful because many ways of analyzing queuing systems produce either L or W but not both. Then equation 1 permits an easy conversion from one to the other of these performance measures. Queues and waiting are ubiquitous in manufacturing: jobs to be done, inventory in process, orders, machines down for repair, and so forth. Therefore, equation 1 finds many uses.

As a mathematical theorem, $L = \lambda W$ has no necessary relationship to the world. Given the appropriate set of mathematical assumptions, $L = \lambda W$ is true. There is no sense going out on the factory floor and collecting data to test it. If the real world application satisfies the assumptions, the result will hold.

The basic tautological nature of the proof can be illustrated by drawing a plot of the number of items in the system versus time, as in Figure 1. The area, A, under the curve represents the total waiting done by the items passing through the system in the time period T. Since the average number of items arriving in a time period T is λT, we have as the average wait per item (at least to first order, with an accuracy that increases as T becomes larger): $W = A/\lambda T$. However, the same area, if divided by the time, also represents the average number of items in the system during the period: $L=A/T$. Eliminating A from these two expressions gives equation 1. Thus, the two sides of equation 1 are really two views of the same thing and, with appropriate treatment of end effects and the taking of mathematical limits, become equal.

Physical laws are different. For example, the equality of the two sides of Newton's law, $F = ma$, cannot be taken for granted. Each must be

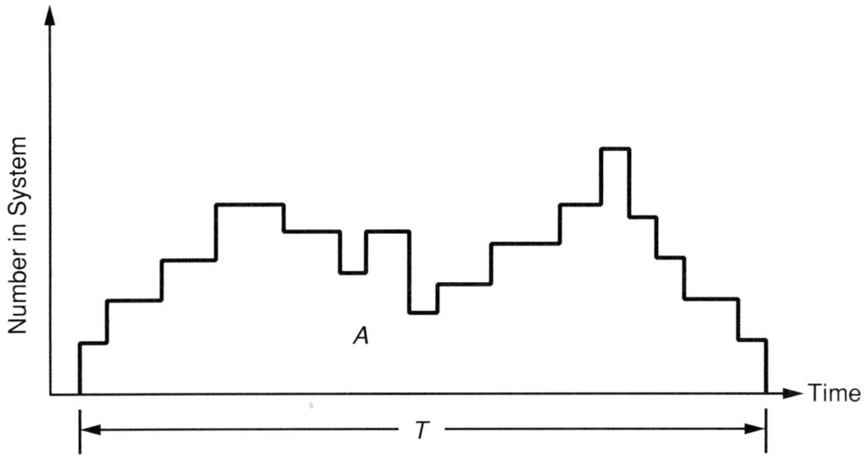

FIGURE 1 The area, A, under the curve represents the total waiting done by the items passing through the system in the time period T.

measured separately and the equivalence verified experimentally. In fact, as is well known, $F = ma$ is only approximate and breaks down at velocities approaching the speed of light. Thus, physical laws require observation of the world and induction about the relationships among observable variables.

LAWS VERSUS EMPIRICAL MODELS

Nineteenth-century physics produced many "laws of nature": Hooke's law, Ohm's law, Newton's laws, the laws of thermodynamics. By the mid-twentieth century, however, many of these laws had been found to be only approximate, and many new, messy phenomena were being examined. As a result, scientists became more cautious in their terminology and started speaking of models of phenomena. This continues to be the popular terminology today. Such is particularly the case in the study of complex systems, social science phenomena, and the management of operations. The word *model* conveys a tentativeness and incompleteness that are often appropriate. We enter a class of descriptions of the world in which there are fewer simple formulas, fewer universal constants, and narrower ranges of application than were achieved in many of the classical "laws of nature."

The incompleteness of models, however, has a virtue in engineering and in applications to managerial decision making. We should include in our models that which is important to the task at hand and leave out that which is not (Little, 1970). This objective is different from the traditional scientific one of describing the world with fidelity and parsimony. For decision-making purposes, we want to restrict ourselves only to the detail required for the job to be done.

Much valuable knowledge can be packaged into empirical models. Their accumulation into organized bodies of learning represents scientific advance and provides a basis for engineering and managerial practice. Here are a couple of examples.

If you examine communications between pairs of individuals in R&D groups versus the physical distance between them, you find a curve like Figure 2 (Allen, 1977, pp. 238-239). Although there is no strictly prescribed functional form or universal constant here, there is definitely a general shape and an experimentally determined range of parameter values. The regularity of the curves can be distorted by a variety of special circumstances, such as electronic mail, location of people on different floors, and the presence of a coffee machine, but the basic phenomenon is strong and its understanding is vital for designing buildings and organizing work teams effectively.

Another example is the *experience curve*, which is illustrated in Figure 3. It is well known that manufacturing costs per unit tend to decrease with cumulative production. This has been documented in a variety of cases

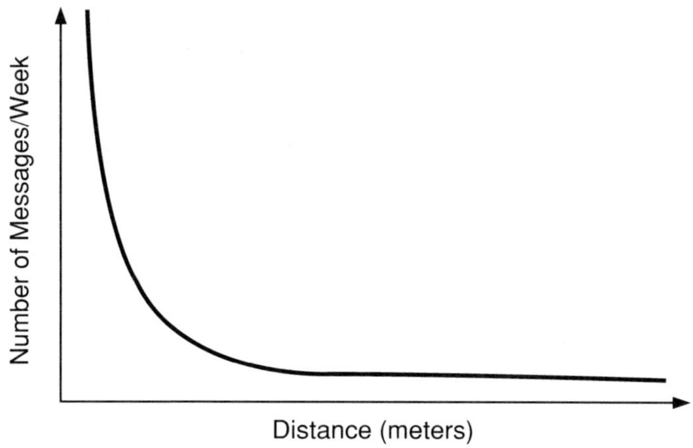

FIGURE 2 Communications between R&D groups as a function of their physical distance from each other.

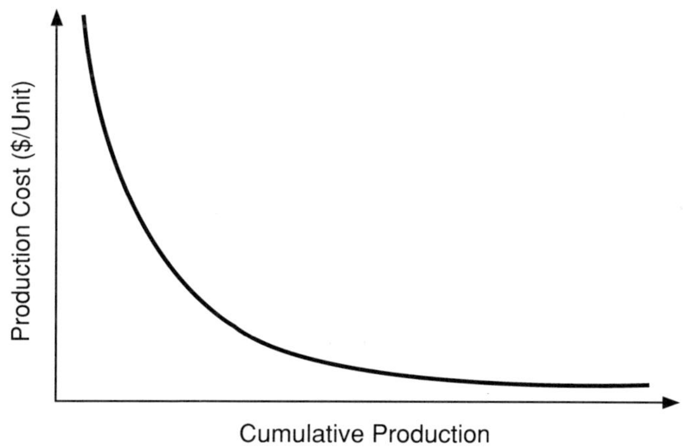

FIGURE 3 Experience curve illustrates the decreasing unit costs for manufacture with accumulated production.

(see, for example, Hax and Majluf, 1984, p. 111). However, the experience curve is a somewhat different kind of relationship than the communications example, because the decreasing cost does not happen automatically. Rather it is the result of much purposeful activity in managing the manufacturing process. In a sense, this seems less satisfying than, say, Newton's law, which predicts unequivocally the trajectory of a ball after it has been struck by a bat.

As another example, even further away from the well-calibrated formulas of nineteenth-century physics, consider *prospect theory* (Tversky and Kahneman, 1981). This describes how people make decisions under uncertainty. As a result of many experiments in which people make choices in different situations with uncertainties, Tversky and Kahneman have produced a descriptive theory of how people make such decisions. They illustrate it with Figure 4, which shows a hypothetical value function for an individual, expressing the person's utility for the outcome of some decision. The curve displays three interesting characteristics of people's behavior. First, people tend to make decisions based on potential gains or losses relative to some reference point. If you change the reference point you are likely to change how they value the possible outcomes of a choice and may therefore affect the choice itself.

For example, if a person has, as a reference point, the catalog price of a particular product and then finds the item in a store at a lower price, he or she is likely to treat the difference as a potential gain. Subsequently, if the person buys the product, the purchase is likely to be considered especially satisfactory, and, in fact, the price "gain" may have helped stimulate the transaction. This is why stores that are running sales usually display the original price prominently. This sets the reference point and makes the discount a net gain for the customer.

A second characteristic of Figure 4 is that the slopes of the curves describing gains and losses are different near the origin. The steeper slope for losses indicates that most people dislike a loss more than they like a corresponding gain. This helps explain the current unfortunate tendency

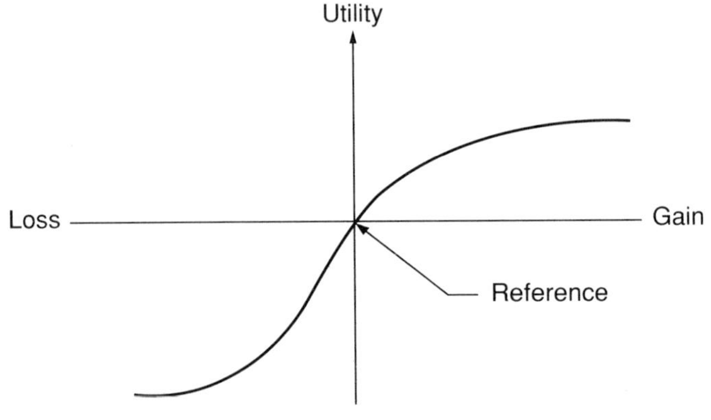

FIGURE 4 Prospect theory illustrates the tendency of people to treat gains and losses differently.

toward negative political advertising. A quantity of negative information suggesting that a candidate might do something harmful if elected may have more influence on the voters than a similar quantity of positive information.

As a third property, Figure 4 indicates that people treat gains and losses differently by showing a concave curve for gains and a convex one for losses. The concavity for gains says, for example, that two separate small rewards to an employee are likely to be appreciated more than a single reward with the same total value. The convexity of losses means that people find it desirable to combine a number of small losses into a large one, as we do when we charge by credit card and pay a monthly total bill instead of several individual ones.

Thus, the shape of the curve in Figure 4 sheds light on a whole variety of phenomena, even though prospect theory does not have the specificity and precision of an $F = ma$. Contemporary psychology is making impressive strides in understanding human behavior, but it often does so more by identifying phenomena and indicating the direction of effects than by producing calibrated models analogous to physical laws.

OUTLOOK FOR LAWS OF MANUFACTURING

What can we anticipate, then, in terms of laws of manufacturing? Are there more laws like $L = \lambda W$? Probably so, in the sense that we should be able to find other useful mathematical relationships that map well onto the world and provide valuable insights about operations.

One example might be duality in linear programming. To any linear program (say, a minimization) there corresponds another linear program (a maximization) that uses the same set of constants but involves a new set of variables. The variables in the new problem have an important operational interpretation in the original one, namely, as prices for changing the constraints. The new linear program also has the fascinating property that the optimal value of its objective function is the same as that of the original problem. Familiarity with linear programming and duality provides a powerful framework for thinking qualitatively about many scheduling and resource allocation problems and for building specific manufacturing models.

Another candidate could be the economic lot size model. Faced with a fixed setup cost, a linear carrying cost of items produced, and a constant sales rate, how many items should be produced? A simple formula provides the answer. In turn, the formula produces insights, such as the result that total costs vary with square root of sales rate.

I am less optimistic about finding many analogs of laws like $F = ma$, because our systems are quite complicated and messy. (Of course, we use the laws of physics directly in the engineering of manufacturing systems.)

On the other hand, I expect much valuable new knowledge to be developed about manufacturing in the form of empirical models like the experience curve. These models follow the spirit of classical physical laws without the three-decimal-place accuracy. Manufacturing is characterized by large, interactive complexes of people and equipment in specific spatial and organizational structures. It seems likely that researchers will find useful macro- and micro-models of many new aspects of these systems. Such models may not often have the cleanness and precision of an $L = \lambda W$ or an $F = ma$, but they can generate important knowledge that designers and managers of manufacturing systems will use in problem solving.

ON COMPLEX SYSTEMS AND THEIR MODELING

Manufacturing systems are often remarkably complicated, involving not only machines and organizations of people, but many and varied information flows and control processes.

Since humans have finite intellectual capacity or "bounded rationality" (Simon, 1957), they like to break systems down into more manageable pieces for analysis, design, and management. This approach is effective, but, of course, the pieces sometimes interact in unexpected ways. Once we have decomposed a system into pieces, we then have a desire to resynthesize the small entities into big ones and work with the large entities as new units. Such hierarchical modeling is certainly a useful approach, if not without its pitfalls.

Large-scale simulations run in computer languages designed for the purpose are now quite common (Pritsker, in this volume). We have outstanding computer capabilities and increasing experience in modeling of complex systems. However, care must always be exercised in order not to lose sight of the forest for the trees. I would argue for having simple models both before and after a large-scale simulation. Before one begins, it is important to ask what are the critical phenomena relevant to the decision or at hand. It can be helpful to build an informal model to represent these phenomena. It is likely that such a model will make too many assumptions to be accepted for the decision at hand, so a more complex and detailed model may be desirable. However, if the results of running a complex model suggest a particular course of action, it is imperative to know why the model produced those results, that is, what were the key assumptions and parameter values that made things come out as they did. We should have a simple model that uses a few key variables to boil down the essence of why the recommendations make sense.

Another approach to thinking about complex systems has been advanced by *system dynamics*. Two separate streams of development can be identified here. One might be called "qualitative thinking through quantitative models." This involves representing by computer variables a variety of interacting quantities, some of which may be quite abstract and not directly measured; examples might be product quality or service level. Other entities could be more concrete, such as production rates and inventories. The goal is to construct a dynamic model in which key variables interact with each other such that the model exhibits the major characteristics of the system's observed behavior in the external world. The results sought from such an analysis are often qualitative, the goal being to understand system behavior: Are there instabilities? What is the system's sensitivity or insensitivity to changes in parameter values?

Other system dynamics models are calibrated on large data bases to represent specific operational systems with statistical fidelity. Some of these models applied to large project management have been very successful, for example, in shipbuilding (Cooper, 1980).

The Need for Multiple Views

The building of more and more complicated models of systems using the same methodologies is likely to yield diminishing returns. Managers face dozens of different problems each day; not just late schedules and excess inventories but also issues such as key people being hired away, roofs that leak, customer dissatisfaction with products, and employee absenteeism. Thus, a hundred different models are often needed, not one big model.

Modeling Myopia

People trained in operations research and management science or in engineering tend to think top-down, that is, in terms of objective functions, control or design variables, models, synthesis of systems from subsystems and the like, with the goal of using the entities under their control to maximize system performance. Consider, however, the following observation by Konosuke Matsushita of Matsushita Electric Industrial company (Stevens, 1989).

> We are going to win and the industrial west is going to lose; there's nothing much you can do about it because the reasons for your failure are within yourselves. Your firms are built on the Taylor model; even worse, so are your heads. With your bosses doing the thinking while the workers wield the screwdrivers, you're convinced deep down that this is the right way to run a business. For you, the essence of management is getting the ideas out of the heads of the bosses and into the hands of labor. We are beyond the Taylor model; business, we know, is now so complex and difficult, the survival of firms so hazardous in an environment increasingly competitive and fraught with danger, that their continued existence depends on the day-to-day mobilization of every ounce of intelligence.

Whether or not Mr. Matsushita's forecast is correct, he forcefully articulates a critical idea—the need for empowering and enhancing the effectiveness of people at all levels of an organization.

We indulge in modeling myopia if we believe as system analysts that we can (or should) be building complete models of our systems and setting all the control variables. Doing so misses the major opportunities for system improvement that are possible by finding new ways to empower the people on the front lines of the organization by giving them information, training, and tools with which to improve their own performance.

Another theme implicit here is that organizational coordination is something much more than top-down control. Interesting new ideas are evolving in this area, for example, developments in computer-assisted collaborative work and coordination theory (Malone, 1988). As information technology has decreased the cost of communication, there has been a growth of lateral communication and coordination and a shift from vertically hierarchical organizations to more lateral and marketlike structures. Lateral coordination is valuable in speeding new product development, finding process improvements, implementing new ideas, and generally facilitating parallel operations in different physical locations.

Thus, in analyzing and designing manufacturing systems, we need to combine new organizational and managerial knowledge with that from physical and operational systems. Many of the issues involved are ill understood today and create fruitful research agendas.

Taking Risks in Manufacturing

DAVID B. MARSING

One of the biggest pitfalls of a company enjoying a particular market and product dominance is the tendency to avoid risk. This is a particularly serious challenge for those companies, because continuous improvements in manufacturing operations are one of the most effective means of creating and maintaining competitive advantage. But improvements in manufacturing operations are typically associated with change, and with any change there is an element of risk. As companies become risk averse, change, except for historically proven methods, is usually minimized although the pressure for increasing output and reducing costs continues.

Products, processes, and manufacturing plants have life cycles, and, when we are lucky, these coincide with external business cycles. Unfortunately this seldom happens. To achieve higher productivity throughout the entire life cycle of a manufacturing operation requires change. Many of the changes needed to prepare an organization for long-term transitions can take years to work through. Limitations on change can cripple a manufacturing company's ability to improve or even take advantage of future business opportunities. Understanding the risks associated with change, preparing to deal with the uncertainties inherent in change, and having the facility to integrate specific changes within a longer-term strategic plan are paramount to the success of a manufacturing company. In the best of circumstances, risks can stall growth if they are incorrectly managed; in the worst case, they can shut down a factory.

RISK AS A CONSEQUENCE OF CHANGE

Change and its accompanying risk can occur in numerous dimensions of the manufacturing operation. Changes to the structure of the enterprise, changes in the people in the various roles of the corporation, and changes to the spirit or attitude of management create organizational risks. On the other hand, changes to manufacturing processes, substitution or selection of new component materials, or increased product performance characteristics at the behest of customers expose the firm to technological risks.

When considering change, the arguments against it can be very persuasive, especially when current performance is successful and competitive pressures are unforeseen. Also, changes to complex systems are more likely to be riddled with unforeseen reactions and unexpected risks. But, for manufacturers that intend to drive markets rather than just react to competitive pressures, the active pursuit of change in product technologies and process capabilities must be recognized as an integral component of long-term strategy. The challenge is to minimize the risk assumed and maximize the benefits gained.

While instances of change seem to be easily classified as organizational or technological, in practice they are not easily separated. Organizational and technological changes are intertwined such that successful change in one category often requires accompanying changes in the other. However, we believe that there are ways to increase the likelihood of successful implementation of change and management of risk within our long-term competitive strategy. These include the effective use of people, the adequate planning for change, and providing of the appropriate tools.

EFFECTIVE USE OF PEOPLE

Total Employee Involvement

For senior managers to sit in their offices and assume that they have all the knowledge and experience needed to make critical risk decisions is a prescription for failure. To make the best decisions with the most knowledge, managers have to use the combined intelligence of the entire work force. But just saying this does not make it happen; to capitalize on the human resource and experience requires empowering the work force. Participation in the strategic planning process ensures that all employees understand the direction in which the organization is moving and why. Everyone in the organization needs to understand his or her role in achieving the factory's goals. And when managed correctly and reinforced regularly, people will actively participate in the execution of the plan (see also Badore, this volume).

Effective Teams

Most companies that have made effective use of their people have done so using teams. Organizing teams of people into functional or strategic groups is a good start. All people in the teams need training in specific areas to help ensure that the team process will work. These areas include how to conduct and participate in meetings, the fundamentals of project management, problem-solving methodology, and conflict resolution.

As teams take on more responsibilities and assume ownership of their areas, the demand for engineering support declines. This becomes an opportunity for reducing the expense of indirect headcount or redeploying these people into tasks that add greater value.

To be able to reduce indirect headcount, individuals and teams on the factory floor must be able to measure the performance of their area and, when problems arise they must be able to diagnose and correct them. Using the concept of continuous improvement, the bulk of the improvement effort in the organization needs to be centered on reducing variability around the process targets. Although manufacturers have been trained and conditioned to "keep things in spec," this is not good enough. Successful global competitors continuously try to tighten the process distribution around their targets, which in most cases are well within the "spec limits." This approach gives each production area a continuously improving process capability. As variability improves and distributions tighten up, there are fewer excursions and thus a need for fewer people to be involved (which will also reduce variability). Using basic statistical process control rules, we can identify which steps are stable and which steps are out of control. These areas are where the technical talent and key teams can have the biggest impact.

Senior management needs to give teams the responsibility and authority to run their areas in both a tactical and a strategic manner. Senior management's role is to understand what the obstacles are and to remove them. Anything less will not build the foundation needed in the organization to deal with change and risk taking.

PLANNING FOR CHANGE

Strategic Planning

Strategic planning is used to create a "planned crisis" that focuses the people and energy on tough goals that are derived from market knowledge and technological opportunities. At Intel we use our manufacturing capabilities as a strategic market force to drive down the cost and increase yields to prolong the life of each of our integrated circuit manufacturing facilities. Because of the rapid changes in the technology of the product and the accompanying production processes, we must constantly strive to reduce our costs and raise our production yield. These planned crises provide the numbers and projections to illustrate the importance of increasing yield while squeezing costs.

Strategic planning is the cornerstone of success in the long run. It is a process that must be iterative and gets as much participation as possible. Setting specific goals and brainstorming on how to get there is the start. The senior management of a manufacturing company needs to understand the ramifications associated with reaching the mature phase in the life cycle of a plant or a manufacturing technology, especially in light of more fierce global competition. In addition it is essential to integrate contingencies to deal with the external influences that can derail an otherwise sound long-term strategic plan. New products or technologies can significantly shorten the life cycle of the production facility. Other influences that need to be considered include economic fluctuations, higher than expected demand for products, change in management (and corresponding change in management style), and even mergers and acquisitions. During this planning process, senior management must continue to ask "What can go wrong?" How would we deal with all of these changes, and how would they affect the assumptions that we have used in our strategic model? How would they affect our execution plan?

Clearly, effective long-term planning requires regular feedback of real changes in the environment, the capability of the factory and even new ideas that come from the organization. The long-term goal seldom changes, but details in executing the plan to reach the goal often will change.

Managing Organizations in Transition

Historically factories have gone through major changes in their business or organization. These decisions are almost always made by senior managers who expect people to get on with their new organization, business, job, or supervisor. Yet we are always amazed at how individuals and teams who worked well before a change slide in their performance afterwards. Management usually fails to consider the human emotions involved in organizational change. Writing on individuals going through change and corporations managing transitions, Bill Bridges (1988) describes three phases people go through during changes or major transitions: an ending, a neutral zone, and a new beginning. His work also describes how this process needs to be allowed to happen for a change to come to fruition. Coworkers and management need to help facilitate this process and to be understanding of people during these transitions. We have to plan for change and to withhold our normal expectations of people or the organization until they get through the emotional transition.

Everyone, to some extent, will go through this transition process, whether it involves a merger, a major reorganization, being assigned to a new team, or even a plant closure. The difficulty is that individual behaviors and reactions differ. Senior managers of manufacturing plants have to consider the changes and risks associated with long-term competition and profitability. They have to plan for the emotional transitions that the work force, on which they depend, will go through. They have to demonstrate leadership and plan their steps.

PROVIDING PROPER TOOLS

Statistics and Problem Solving

Teams organized to run and improve their areas in a manufacturing operation must acquire the appropriate capabilities in statistical methods and problem-solving techniques. Developing a working knowledge of these methods is essential in monitoring the performance of an area and being able to identify the parameters that afford the highest leverage for improvement. As this capability evolves, there is a corresponding opportunity to implement a process control system used and maintained by production teams and individuals.

The benefits of using such a system are that it promotes teamwork, systematically empowers direct labor production people, improves the learning rate, and improves material quality and product yields. Communications between departments and across work shifts is drastically improved with the use of a common process control system. Properly set up, such a system shares knowledge through documentation. It provides a logical and consistent game plan to deal with variability in the production process and equipment performance. Holes in the existing knowledge quickly become apparent. Engineers spend more time improving the production processes and less time doing routine tasks (such as fixing equipment or adjusting machines to get them back in tolerance).

At Intel we have a mechanism for transferring responsibility for management and control of the production process to the operators of equipment and from the process engineers. This mechanism has supported our strategy to force the costs of the production process down the cost curves. A side benefit, in many instances, has been to stabilize the production process. Our mechanism is a combination of statistical process control (SPC) charts that are used to manage each production process step and response flow checklists (RFCs) to troubleshoot and correct problems. SPC is applied in the usual manner to data collected and evaluated by the equipment operators. The RFC is Intel's way of transferring, to the operators of the equipment, the knowledge and experience accumulated by the process and equipment engineers as they installed and debugged the production machinery. The information in the RFCs is a series of structured "if then. . . else. . ." statements that lead an operator through a well-defined sequence of diagnostic questions and associated corrective actions or adjustments.

Change Control Methodology

As people in the factory see opportunities to make improvements, there will be increasing pressure to implement changes. In an integrated production process, seemingly harmless improvements can throw a downstream processing step totally out of control. Understanding, through well-designed experiments and characterization, of the interdependencies between processing steps is critical to minimize risk in the improvement process. To help manage this kind of effort, it is useful to organize a change control system. This system should use a structured approach to experimental design, agreed upon criteria for interpreting results, and the confidence level required to implement a change. Customer's issues need to be considered. When changes in the process or product are considered major, qualification is usually required from the customer.

To maintain consistency with a structured change control procedure requires educating the work force in the philosophy of change control and the mechanics of the process. The committee of people who review these proposals and reports need to represent most of the disciplines in the factory and should be senior managers. The membership of these committees need to be consistent. It is important to ensure that this kind of a system is perceived not as an obstacle to change but rather as a methodology for assessing risk and making decisions in a consistent manner.

In an Intel study of all changes intended to improve factory yields, only 18 percent actually resulted in a positive yield improvement.[1] Yet, in each Intel factory there were change control systems similar to the structure previously discussed. Although many changes were intended to improve manufacturability, the data suggest that the number of interdependencies in the process steps were not well enough understood. We discovered that in a high-yielding, complex manufacturing process for semiconductors, there are no home runs or major breakthrough changes. The biggest gain comes in smaller continuous improvements.

---

[1]In the manufacture of integrated circuits, yield is measured at several points during the production process: the line yield reflects the scrap rate for entire wafers, while yield-to-die measures the number of good die obtained from each silicon wafer.

SUMMARY

Many people express the concern that manufacturing in this country must change, or else our ability to compete and influence industrial and technological growth will disappear. One of the cultural and social advantages that American people have is the ability to take emotional risks in our organizations, absorb changes in a fluid and nondisruptive manner, and deal with evolving group structures and relationships. For manufacturing this is an asset that may indeed be a globally competitive advantage for the United States.

Our ability to capitalize on this asset, however, will require us to develop an excellent working knowledge of change management and the risks associated with it. We must fundamentally break the paradigms established by our education and the structural norms of classical American manufacturing, which have conditioned, and now impede, our competitive opportunities. Senior management must learn how to empower their work force, and they must begin to assume the role of leaders. Employees must be given the knowledge and training to be effective team players and management must establish a culture to promote this type of organizational behavior. As the work force is empowered to take ownership of its areas of the production process, it must also have the tools to understand and improve factory performance. To manage the changes associated with improvement requires an established methodology to review proposed changes, establish the criteria for success, and assess the attendant risks. It is essential that everyone understand and support a long-term goal to be competitive, even if it is not at all clear how to get there. Establishing a strategic planning process will help build a foundation for the factory. This planning process must be iterative and open to all ideas, and it must leverage the strengths and capabilities of the people, plant, and production process.

Once the direction is set and the systems are in place to allow a factory to design change, take risk, and position itself to be competitive, it is critical to establish an environment that reinforces the notion that it is good to take risks. Consistency in management's attitude toward change and evolution of the organization to deal with the requirements is paramount.

Constant Change, Constant Challenge

JOE H. MIZE

Historically, "factory modernization" initiatives were singular events occurring once every 20 to 25 years. The factory structure was essentially fixed for long periods of time, and product life cycles were much longer than they are today. Relatively small capital expenditures were required between major modernization projects. The same was true for the "systems," policies, and procedures of the organization.

Today's environment requires a very different approach. Companies can no longer use equipment until it "wears out"; it will become obsolete long before. Consequently, factory modernization today must be considered a continuous, ongoing process in which 25 to 30 percent of all processes and systems are being replaced annually in many industries.

Managers are rapidly losing many of the planning aids that have allowed them to proceed in an orderly, progressive fashion. In the past, managers could safely assume that tomorrow will be much like today, with only marginal changes. In fact, randomness was often much larger than the average marginal change; thus, the "noise" masked the "signal." Consequently, many of today's managers know how to manage only on the margin, in a static mode.

Today's managers are faced with the fact that change is continuous, pervasive, and often traumatic. It affects every level and every aspect of the organization. A rapidly changing total environment has become the norm, replacing the relatively stable and static environment of the past.

NEW MANAGEMENT PARADIGM NEEDED

Traditional organizational structures, management practices, and organizational policies are proving inadequate for the new environment. It appears that a complete paradigm shift will be required in order to cope with the new challenges. The new management mind-set must be based on the realization that rapid, continuous change is the norm, not the exception. At this point, we have only vague notions and tentative hypotheses regarding the nature of the new management paradigm. Table 1 contrasts the traditional and change-driven management environments. As Davis has observed (1987, p. 8), an important aspect of any management paradigm is the prevalent view of time:

> In the industrial economy, our models helped us to manage aftermath, the consequences of events that had already happened. In this new economy, however, we must learn to manage the beforemath; that is, the consequences of events that have not yet occurred.

Increasingly, managers will have to visualize their businesses and organizations at a point in the future, interpolate their way backward into the current reality, and then aggressively manage the implementation of the transition path from here to there. But the future vision is a moving target, and the backward interpolation process must be ongoing and dynamic.

MANAGING FROM THE FUTURE

Managers of organizations must learn to lead and to create plans relative to a future point in time, with a mind-set that assumes that one has already arrived at that point in time. This is possible due to the unique human ability to project oneself into the future and to create in one's mind a desired state of affairs. This mental act may be called "visioning," and almost everyone can do it to a certain degree.

Far from everyone, however, can effectively execute the next step in converting visions or dreams into tangible results. We are speaking of the ability to work backward from the desired future state to the present state in a way that clearly and unambiguously delineates an achievable action path. Most of us have great difficulty in dealing with dozens of interdependent variables, sorting out the complex cause-effect relationships, and relating all the dynamically changing outcomes to explicit decision variables over which we can exercise day-to-day control.

The rare individuals who have this capability seem to be blessed with a unique set of mental processes.

> Their intuition is generally ahead of their conceptual framework, and they evolve a coherent and post-facto rationale for the details of what they are

TABLE 1 Traditional Versus Change-Driven Management Environments

Factor	Traditional Environment	Change-driven Environment
Time Factors		
Long range	5 years	1 to 2 years
Medium range	2 to 3 years	6 to 18 months
Short range	3 months	1 week
Frozen master schedule	6 months	1 month
Frozen operational schedule	1 month	1 day
Machine schedule updates	1 day to 1 week	continuous
Facility modification	2 to 4 years	1 month
Equipment Replacement	1. Replace when it wears out 2. Avg. 8% per year, capital equipment replacement	1. Replace when it is obsolete 2. Avg. 30% per year
Prototyping, Scale-up, Full Production	1. Done off-line, with transition to full production requiring 3 to 9 months 2. High rejects initially expected 3. Tooling designed during transition	1. Performed primarily in simulations, going directly to production 2. Parts-per-million quality level from day 1 3. Tooling designed before conversion
Training	On-the-job training, irregular and infrequent	Professional, continuous
Job Design	1. Very narrow boundaries 2. By industrial engineers, off-line	1. Broader classifications 2. By workers, on-line
Quality	1. Quality control charts designed by QC 2. Inspectors, off-line, after the fact 3. Send rejects to rework off-line	1. Produced from CAD data base 2. Operator, on-line, real time 3. Shut down line, fix problem
Work Flow	Sequential	Parallel
Planning	Reactive	Responsive, learning, anticipatory

> already doing. For those whose strategy flows from their actions, rather than vice versa, strategy is the codification of what has already taken place; it is the writing of future history (Davis, 1987, p. 27).

This describes the mind-set of many entrepreneurs, and suggests that organizational strategy formulation based on insightful intuition is superior to that based on traditional formal planning.

The challenge we face is to translate "insightful intuition" into a logical and conscious process amenable to organizational implementation. This process must be understandable and executable by the large majority of managers who do *not* possess the unique intuitive abilities described above.

STRATEGIC CONTROL: THE LEARNING ORGANIZATION

The early practice of strategic planning amounted to little more than extrapolation of the past into the future. Most formal planning processes attempt to characterize the future several years out and then monitor the organization's progress toward the vision on a rolling basis.

Figure 1 shows the basics of traditional planning processes. In this model of management planning and execution, the control system monitors past outcomes, usually on a quarterly basis. Only minimal learning results from this approach. Since there is no formal linkage back to the planning

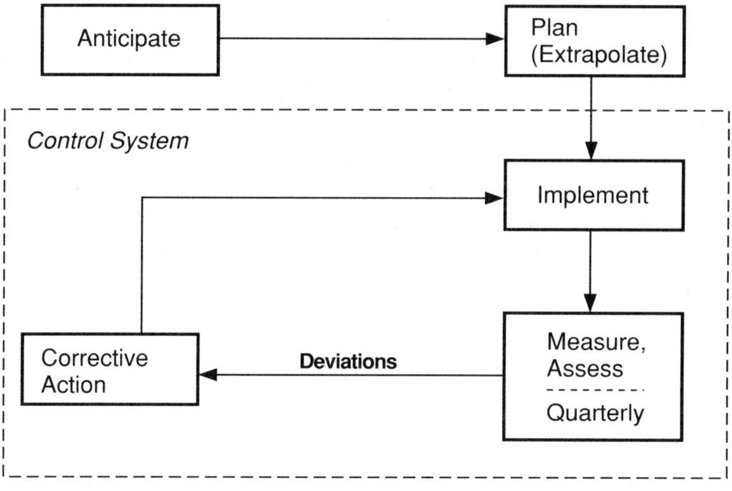

FIGURE 1 Traditional formal planning process: An open loop and minimal learning.

function, the system operates essentially in an open-loop mode with respect to the planning function.

The weaknesses of this traditional process underscore the need for a conceptual framework for planning and managing that incorporates visioning, futures evaluation, and organizational learning. As suggested by Davis (1987), the tracking portion of the control function should be placed in the visioning component of the model, so that it continuously tracks how the projected future is changing and determines the best strategies for the organization to follow. Although not mentioned by Davis, a control function is still required at the operational level.

Figure 2 illustrates the conceptual framework for a learning organization. In this model, the control system operates at two levels. First, it monitors a simulation of the future iteratively until an acceptable organizational strategy has been identified consistent with the vision of the desired future state. In a sense, this control structure is a feed-forward control loop.

Second, a feedback control loop tracks actual results, compares them with the planned results emanating from the organization strategy, and determines appropriate corrective action relative to operational performance. It is important to note that this model captures corporate experience and imbeds the "knowledge" accumulated from strategic and operational experience in the "corporate memory" for use in future planning.

The strategic control model shown in Figure 2 may be considered as two highly interrelated subsystems acting as an integrated whole. The dashed lines indicate the composition of the two subsystems. Some of the terms included in the diagram are defined as follows:

Competitive benchmarking: A process of systematically assessing critical performance attributes of a firm and comparing the firm to the best comparable firms in the world, relative to those attributes.

Entrepreneurial surveillance: The practice of aggressively surveying the total business environment, with the goal of identifying all pertinent opportunities for enhancing competitive advantage.

Model calibration: The process of continually "fine tuning" parameters within the Corporate Strategic Model, in light of experience gained in operations.

Knowledge base: The organization of data and information into rule sets for use in creating inferences relative to strategic choices that may be made.

Visioning: Strategic posturing with respect to possible future scenarios.

Organization strategy: The collective and comprehensive intention of the firm.

Adaptive control: The process of capturing experience to improve the comprehensive closed-loop system of planning, execution, and control.

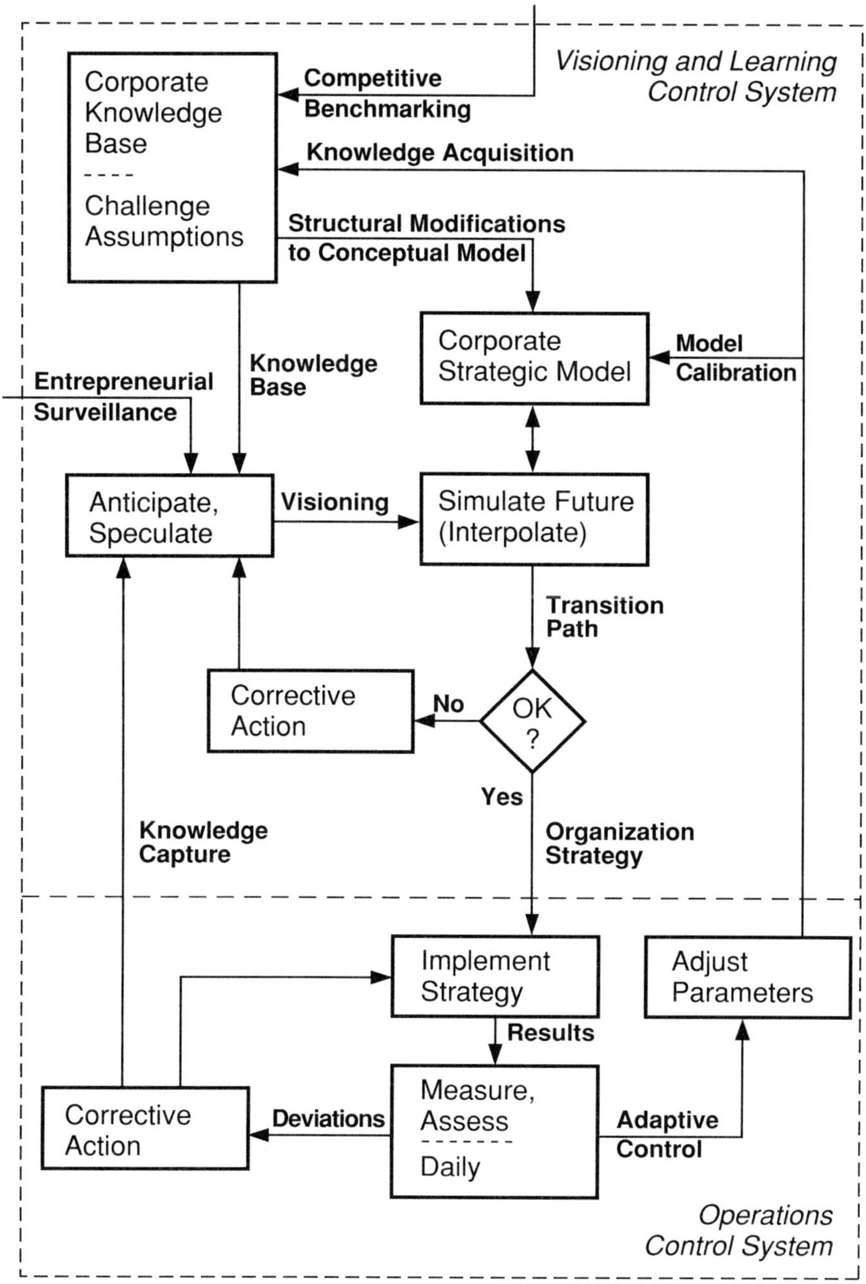

FIGURE 2 Strategic control model: Continuous learning and improvement.

A significant advantage of this model is that it accommodates equally well the concepts of those who advocate "breakthrough planning" as well as those who advocate "continuous improvement." These concepts need not be in conflict. This model of organizational control facilitates continuous learning and improvement through the application of new knowledge. It also facilitates, when appropriate, the kind of broad, sweeping change that results in order-of-magnitude improvement in a short period of time.

ORGANIZATIONAL LEARNING

A major element of this suggested framework is "organizational learning," a concept that is not well understood. There are essentially no tools proven useful for performing this essential function. Many of the decisions that must be made in an organization are either similar or identical to decisions made in the past. Yet, because these past decisions and their results were not recorded, new decisions are based on whatever is remembered from the past and on the perceived results of past decisions. Because human beings are not able to remember details of hundreds of prior decisions, let alone the consequences of those decisions, there is a critical need for research in "corporate memory processes."

Another shortcoming in available methodology is evident from a close examination of the strategic control model depicted in Figure 2. Our ability to "simulate the future" is primitive. Current modeling and simulation methods are grossly lacking in this regard, and research should be directed toward filling this void (see also Mize and Beaumariage, 1988, for related research needs).

Until better tools and methodologies are developed, organizations have little choice but to rely on human expertise to perform the learning functions described above. In fact, humans will *always* be responsible for these functions, even when better tools and methodologies are available. The tools will be decision aids; they will not make the decisions.

Hayes, Wheelwright, and Clark (1988, pp. 268-269) make several important points regarding organizational learning:

- "The two essential tasks of management are to create clarity and order (eliminate confusion), and to facilitate learning."
- "Change is not synonymous with confusion. . . . In a confused environment it is very difficult to determine cause and effect."
- "Reducing confusion and enhancing learning are not contradictory imperatives. To the contrary, they are closely related and powerful in combination."
- "The effectiveness with which people work together . . . has an important influence on the organization's ability to create new knowledge

> and apply it in the production process. Moreover, each time an individual or team completes a learning cycle, they expand their knowledge of the process and their skill in solving problems."

Thus, we see that an organization is truly capable of learning.

SUMMARY

Effective management in today's environment presents formidable challenges. The environment is unpredictable, unstable, and increasingly competitive. Reactive management is no longer viable. Managers who survive and thrive will be those who develop the ability to manage their organizations as a combination feedback and feed-forward control system. Central to this ability will be three critical characteristics:

1. The ability to create a realistic vision of the future state of affairs.
2. The ability to simulate the future environment through the generation and evaluation of strategic scenarios.

Manufacturing Capacity Management Through Modeling and Simulation

A. ALAN B. PRITSKER

Computer-integrated manufacturing (CIM) has been established as an architecture for today's factory and an absolute necessity for the factory of the future. Emphasis in CIM has been on automation, advanced machining capabilities, new organizational structures for facilities and personnel, and information acquisition, storage, transfer, and use. This paper examines CIM from the perspective of the integration of functions associated with management of manufacturing capacity. The premise of the paper is that manufacturing operations should be driven by capacity considerations, not material availability. The manufacturing enterprise must have proven techniques for managing capacity: total capacity management (TCM) including capacity planning and design, finite capacity scheduling, capacity control, and the continuous measurement of total available capacity and its use. Total capacity management is a vital foundation to a corporation seeking to achieve a competitive edge and superior productivity. One of the main goals of the CIM architecture is to provide for capacity management. This paper advocates simulation as the primary means for achieving total capacity management. It proposes the use of a common modeling language and common data to support simulation analyses across the many tasks related to TCM.

MODELS, MODELING, AND SIMULATION

Models are descriptions of systems. In the physical sciences, models are usually developed based on theoretical laws and principles. The models may be scaled physical objects (iconic models), mathematical equations and relations (abstract models), or graphical representations (visual models). The usefulness of models has been demonstrated in describing, designing, and analyzing systems. Model building is a complex process and in most fields involves both inductive and deductive reasoning. The modeling of a system is made easier if (1) physical laws are available to describe the system; (2) a pictorial or graphical representation can be made of the system; and (3) the uncertainty in system inputs, components, and outputs is quantifiable.

Because of the complexity of manufacturing systems, a model builder must decide on the elements of the system to include in the model. To make such decisions, a purpose for model building must be established. Typically, a purpose for modeling is related to a stated manufacturing problem or project goal, which helps set the boundaries of the manufacturing system and the level of manufacturing detail necessary to solve the stated problem. The modeling of a manufacturing system is sometimes difficult for one or more of the following reasons: (1) there is a lack of fundamental physical laws (see Little, in this volume); (2) many of the procedural elements are difficult to describe and represent; (3) the required policy inputs are hard to quantify; (4) random components are significant elements; and (5) human decision making is an integral part of manufacturing operations. The last decade has seen a tremendous increase in the modeling and simulation of manufacturing systems. This can be attributed to recognition of the need to improve manufacturing operations, and recognition that the impact of decisions need to be assessed before the decisions are implemented. The availability of simulation languages to build and analyze manufacturing models has stimulated this growth. Another contributing factor is the availability of knowledgeable industrial engineers who have a simulation language background (Pritsker, 1986a).

As Simon (1990) points out: "Modeling is a principal—perhaps the primary—tool for studying the behavior of large complex systems. . . . When we model systems, we are usually (not always) interested in their dynamic behavior. Typically, we place our model at some initial point in phase space and watch it mark out a path through the future." Manufacturing models analyzed by simulation (simulation models) are developed to study the dynamics of the manufacturing system. Such models are built without having to fit the manufacturing system into a preconceived model structure because the analysis is performed by playing out the logic and relationships included in the model. For this reason, simulation models can be built at either an aggregate or a detailed level. Of fundamental importance is the building of simulation models iteratively, allowing them to be embellished through simple and direct additions.

TOTAL CAPACITY MANAGEMENT: AN OVERVIEW

Simulation has been used to support many different manufacturing activities, including product design, process design, facility design, operational scheduling, and schedule management (Pritsker, 1990). Fundamentally, models developed for simulation analysis relate to the setting of capacity requirements for the manufacturing facility and the determination of how to use the capacity to process orders through the facility. Simulation is further used to manage these activities over time to achieve continuous improvements in manufacturing capabilities.

Figure 1 presents a schematic of the manufacturing production-scheduling-operations environment. Capacity management using simulation involves six functions, indicated by the six shaded blocks of Figure 1: Design assessment; Capacity requirements planning and analysis; Scheduling; Schedule management; Schedule execution and dispatching; and Status presentations and statistics. The functions of master scheduling and production control/MRP II [manufacturing resource planning] can, for some manufacturing systems, be performed using simulation. For this paper, no assumption is necessary regarding the need to perform these two functions or whether simulation is used directly or indirectly to accomplish the functions.

Design assessment involves the use of a model of manufacturing operations to estimate the performance of the manufacturing system for different levels of demand in conjunction with designed or actual process plans and resource allocations. The process plans are part of the model and specify the job steps, including resource requirements, to make the product. A separate model is sometimes developed to characterize the orders that make up future demand. Capacity requirements planning and analysis seeks to determine whether manufacturing operations can process the shop orders released from production control/MRP II in a timely manner. Before detailed scheduling can be done, a finite capacity analysis determines the level of resources required to meet current demand. When capacity levels are set, detailed scheduling can be accomplished by using the model to simulate allocation of available resources at specified start times to the actual jobs included in the shop orders. Since the model contains the detailed process plans or job steps, the start and completion times of each operation can be established, and hence the order can be scheduled. These schedules can then be distributed for schedule management, which entails the use of current operational status and critical issues to adjust the schedule. Maintaining shop floor discipline when adjustments are made is important. The outputs of schedule management are dispatch lists detailing the scheduled time to perform each job and prescribing the required resources. In addition, methods for improving the scheduling process through the collection of data and the parameterizing of rules to improve the scheduling

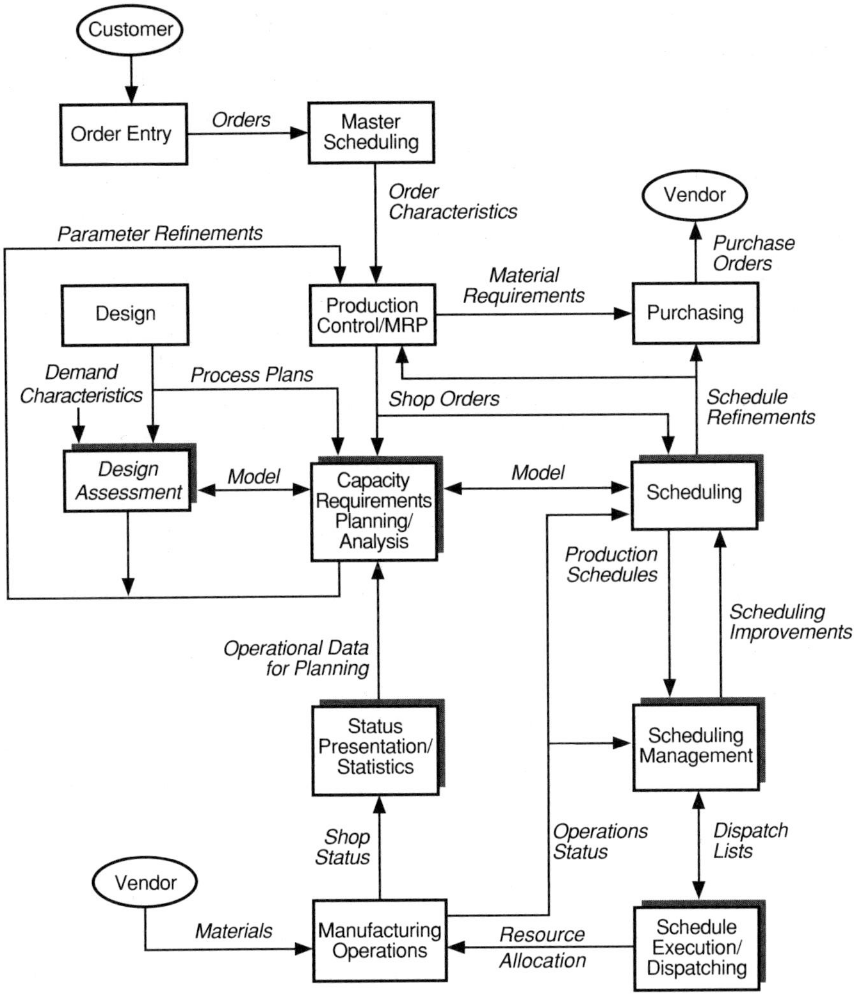

FIGURE 1 Total capacity management description.

process is part of TCM. For example, the application of artificial intelligence tools in conjunction with simulation models can lead to better scheduling practices.

The dispatch lists are the basis for schedule execution and dispatching, that is, the actual resource allocations to jobs. Data on operational status are fed back to scheduling and schedule management to determine the frequency with which new schedules need to be prepared. The display of this status information provides a basis for ongoing decision making. The current status and an up-to-date analysis of immediate past performance can be used in capacity requirements planning where additional simulations can be performed to assess future performance. Through this feedback link, continuous improvements in manufacturing operations can be made and information gathered for future design assessments and new scheduling algorithms.

The feasibility of TCM relies heavily on the ability to build on existing data and models. The use of a common simulation language to obtain a common basis for modeling across the functional problems of TCM makes the evolutionary problem solving described above plausible. It allows for "going to the gemba" (Welliver, in this volume), where gemba is someone else's functional area in which models have been built, data collected, and analyses performed based on the needs of a different functional area.

Models contain information about manufacturing processes, and by using such models continually, the processes will be better understood. *Understanding leads to improved manufacturing and information for improving design.* Thus, TCM is a mechanism to achieve, using simulation, a new form of *Kaizen* (Imai, 1986) by which the processes of manufacturing and decision making can be continually evaluated, changed, and improved. The need for such a mechanism is described in detail in *Dynamic Manufacturing* (Hayes et al., 1988). Innovation also is enhanced, because a model developed in one functional area can be used to indicate the possibility of new constructs for another functional area. Thus, improvement cycles in a single functional area may be used to foster new models and concepts in other functional areas. *The common model, common data foundation* presented for TCM, when fully implemented, *provides a basis for achieving world-class manufacturing.*

FUNCTIONS IN TOTAL CAPACITY MANAGEMENT

In developing an architecture to support TCM, the methods by which TCM functions are performed are extremely important. TCM functions are performed repetitively to achieve continuous manufacturing improvements. Thus, the sequence in which they are performed or discussed is of minor concern.

Design Assessment

Simulation has had its most extensive use in the assessment of manufacturing designs where comparisons of different facility organizations (group technology cells, transfer lines, job shops, etc.) and resource capabilities are evaluated. There is no need in this paper to present a catalog of simulation applications for design problems. Because simulation has been used at many levels across a wide spectrum of systems designs, many types of outputs and analysis capabilities are associated with simulation models. To illustrate this variety of model uses, the primary simulation outputs associated with different levels of model use are given in Table 1. Of course, any simulation output could be employed at any level.

Capacity Requirements Planning and Analysis

Capacity requirements planning entails evaluating the ability of current resource levels to meet current orders and projected demand. The current shop floor status and inventory levels are considered and process plans are used to calculate the load at work centers. In the planning stage, the load at each work center is evaluated with regard to the actual capacity of the work center. Corrective actions are made as required by rescheduling orders, hiring and layoff reassignments, overtime, outsourcing, alternate routing, tooling changes, and so on.

Capacity requirements analysis is concerned with controlling capacity

TABLE 1 Primary Simulation Outputs for Different Model Use.

Model Use	Primary Simulation Outputs
Explanatory Device	*Animations*
Communication Vehicle	*Animations, plots, pie charts*
Analysis Tool	*Tabulations, statistical estimators, statistical graphs, and sensitivity plots*
Design Assessor	*Statistical estimators, summary statistics, and ranking and selection procedures*
Scheduler	*Tabular schedules, Gantt charts, and resource plots*
Control Mechanism	*Tabular outputs, animations, and resource plots*
Training Tool	*Animations, event traces, statistical estimators, and summary statistics*

during the execution of the production plan. It includes the use of models to evaluate the various types of proposed corrective actions. The performance of capacity requirements planning and analysis leads to the requirement for additional design assessments and provides inputs to the procedures used for scheduling jobs. In some cases, capacity requirements analysis is used to set due dates for use in scheduling, this helps synchronize material purchasing and distribution functions with operational requirements.

Scheduling

Scheduling means establishing job start and completion times for the orders that have been released to the shop floor. Scheduling must account for all the specific operational constraints of the manufacturing facility, including limited resources, breaks, shifts, machine availability, personnel availability, material availability, and material handling capabilities. The operational procedures of just-in-time, *kanban*, scheduling by due dates and priority assessment are all included in the computations to produce the schedule. Because of this complexity and the diverse nature of scheduling philosophy, that is, backward scheduling, forward scheduling, scheduling the bottlenecks first, or local dispatching using global information values, simulation is necessary in all but the simplest manufacturing environments. Supporting the requirement for simulation is the complexity of logical conditions based on precedence requirements, constraints, resource availability, material supplies, and personnel contention involved in most manufacturing operations. Optimization, when used, involves the local application of mathematical programming to schedule a subset of orders on a subset of resources. Simulation is usually required to assess the feasibility of the schedules produced by an optimization technique.

Schedule Management

Schedule management entails assessing schedules and the ability to change or manage them. This function will most likely be performed in a graphical and interactive environment using displays based on Gantt charting techniques. Schedule adjustments are made by sliding, interchanging, inserting, and deleting jobs. There is a need for display capabilities that depict job-precedence constraints, resource-use diagrams, and order start and completion indicators. Net change information from schedule management decisions can be analyzed using artificial intelligence techniques to assess the value of new procedures for producing schedules.

Schedule management can also be used to evaluate the effect of expediting jobs or taking on new sales opportunities within the current status and schedule. For this purpose, operational status must be made available to the schedule management function. Evaluating how jobs are performed versus how they were scheduled is part of schedule management. Artificial intelligence techniques could also be used here to determine the measures of current status that indicate whether a rescheduling will produce a significantly different schedule from the one currently distributed to the shop floor.

Status Presentation and Statistics

Data obtained from manufacturing operations can be displayed on a diagram of the facilities on which the operations are performed. By maintaining records of status and status changes, a computer-generated animation of manufacturing operations over a preceding time interval can be shown on a computer screen. Statistics on past operations could be used to answer questions about methods of operation and the order book that drives manufacturing operations. Given the current status of a manufacturing facility and the current orders in process, simulations can be performed to determine the impact of releasing additional orders to the shop floor. These animations can address various "what if" questions to study and evaluate manufacturing operations in a "pretend" mode (Clark and Withers, 1989). Outputs from these simulations provide information for capacity requirements planning and for design reassessments leading to operational improvements. As discussed by Mize (in this volume), simulations of this type provide a feed forward control loop for improving organizational strategy.

Schedule Execution and Dispatching

Schedule execution and dispatching is a function not normally included in capacity management but is included here because of its importance to the philosophy embodied by TCM. If a schedule is not executed as prescribed, the total TCM function will suffer. *A disciplined shop floor is required in which local decisions are not changed without approval or, at least without communication back to the schedule manager. Shop floor personnel must be integrated into TCM activities and be knowledgeable about the process of performing TCM.* Training in the use of models and simulation can provide information to the shop floor operators and a means by which they can provide feedback to improve TCM. With training materials that use the same model and data as the other functions of TCM, shop floor workers can gain a perspective of overall operations that will help them to achieve the goals of the manufacturing system.

TCM ARCHITECTURE

Total capacity management will be performed in a heterogenous computing and software applications environment. MRP II systems, purchasing systems, process plans, and shop floor control will most likely be performed on one or more computers using different data base systems for their individual performance (Baudin, 1990). This will require an integrated architecture for software developments to achieve TCM.

A key to obtaining TCM will be the use of a common modeling language and common data throughout the functions depicted in Figure 1. An architecture based on this concept for TCM is shown in Figure 2. The architecture is layered to include user interfaces, underlying utilities for

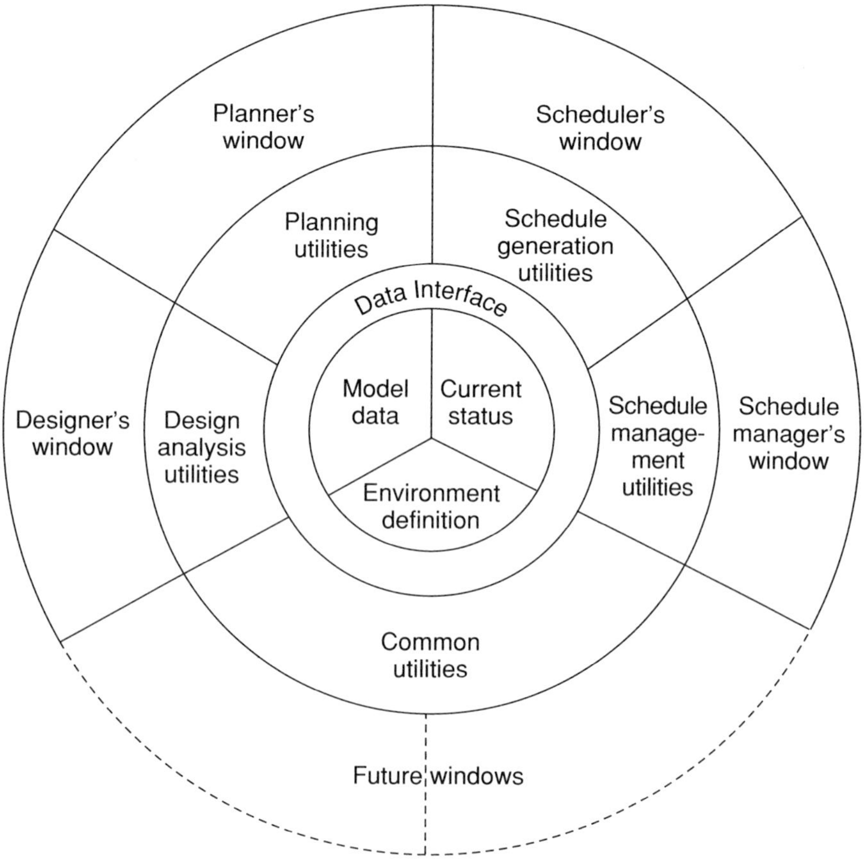

FIGURE 2 Total capacity management architecture.

accessing data through a standardized data interface, and a fundamental reliance on common models and common data storage (International Business Machines, 1989). In this architecture, the user interface is provided through four windows for designers, planners, schedulers, and schedule managers. Each of the windows should have a similar look and feel and be organized to satisfy specific user needs. Although the design of these windows will depend on particular applications, there will be a large overlap in the displays. Figure 3 lists several of the capabilities required for each functional window type. Future windows will be required for direct delivery of information to decision makers and corporate executives. Before this can be accomplished, the roles of the decision maker or corporate executive relative to TCM will need further clarification. If the adaptive and improvement features shown in Figure 1 are performed well, it is conceivable to automate the functions of the scheduler and schedule manager.

The utilities layer in Figure 2 will need to include capabilities for performing simulations, graphic utilities, artificial intelligence, expert system rule building, and interfaces to data bases for accessing information on process plans, orders, equipment characteristics, operational data, other modeling tools, and current status. Other utilities required relate to model building, display generation, animation generation, schedule distribution, and communications in general.

Designer's Window	1 *Graphical builders* 2 *Statistical analysis tools* 3 *Cost analysis/spreadsheet tools* 4 *Animation builder*
Planner's Window	1 *Capacity planning displays* 2 *Resource availability builder* 3 *Shift scheduling builder* 4 *Order manipulation tools*
Scheduler's Window	1 *Multiple schedule reports* 2 *Rules for scheduling control* 3 *Alternative builders for:* a. *Resource schedules* b. *Work order releases* c. *Load manipulation*
Schedule Manager's Window	1 *Interactive Gantt charting* 2 *Expedite function evaluator* 3 *Schedule distribution capabilities*

FIGURE 3 The user interface layer.

DISCUSSION

The concept of total capacity management presented in this paper specifies an integrated set of capacity-related functions to be performed using a common modeling language and common data. It is not a revolutionary approach to the problem. It builds on existing systems and existing data bases. *In manufacturing, an evolutionary process has been shown to have the greatest chance of having a significant effect.* Thus, TCM does not replace production control systems, process planning systems, CAD/CAM systems, or quality improvement systems. *A foundation of manufacturing systems, TCM advocates the integration of functions relating to capacity management and the sharing of information and decisions with those systems that are also involved in manufacturing system improvement.* TCM is focused on the operational capabilities and operations of the manufacturing system. It performs functions related to capacity setting and resource and job scheduling originally promised by MRP II but not currently provided in an accurate or usable form at the shop floor level. It provides a path to break down the barriers between the functional units of design, planning, operations, and control.

TCM as presented here includes self-improving mechanisms for its own operations and the operations of those systems that use the same data sources. TCM as a concept requires further definition, exploration, and design. However, integrating capacity management functions and the development of a system to achieve TCM are feasible using current hardware, software, and human capabilities.

ACKNOWLEDGMENTS

The material presented in this paper is based on past research on simulation languages, modeling languages, development of decision support systems for manufacturing, and many years of applications. Many discussions have been held with colleagues and students concerning the use of simulation for capacity management. In particular, the author would like to acknowledge discussions with David Yancey, David Wortman, Steven Duket, Bill Schaefer, Doug MacFarland, Bill Lilegdon, and Dan Murphy of Pritsker Corporation, whose research and development activities are included in the concepts contained in this paper, and the work with the NAE Committee on Foundations of Manufacturing Systems with Kent Bowen, Harry Cook, Dale Compton, Jim Lardner, and Dick Wilson. This material is also based on discussions with Bruce Schmeiser and Jim Wilson of Purdue University on modeling and simulation research supported by the National Science Foundation under Grant No. DMS-8717799.

The Power of Simple Models in Manufacturing

JAMES J. SOLBERG

Despite being admonished repeatedly to "keep it simple," most of us who deal with technical issues in manufacturing persist in developing evermore complicated models. There are some good reasons (and also a few bad ones) for doing so. At the heart of the matter is the fundamental reality that manufacturing processes are complex. As we advance in our understanding of these processes, it seems natural to incorporate additional complexity in models. Another contributing factor is the increasing power and availability of computers. With this increasing capability comes the opportunity to compute what we never before could compute, and this opportunity presents compelling temptations to see how much we can do. Furthermore, we tend to associate the quality of a model with the degree to which it faithfully represents the system and conditions it purports to model. If we are forced for whatever reason to make simplifying assumptions or to omit details, we regret the necessity and view the step as a loss, a deficiency, a compromise between what we would like to do and what we are able to do. Of course, we know that models always involve some degree of simplification, but we are generally reluctant to introduce more than is forced upon us.

I will call this concern for the faithful correspondence of a model to its referent a concern for validity. The thesis of this paper is that we—those of us in the research community who have been developing models for manufacturing—have been preoccupied with validity to the point that other important attributes of models have been neglected. In particular, we have sacrificed clarity, generality, and (most important) credibility. The ultimate consequence of this imbalance is that many of our models are not used.

Some cynical observers have gone so far as to accuse academics of deliberately introducing obscurity into their papers in the interest of preserving the mysteries of their priesthood. Others have suggested that the "publish or perish" pressure and all that goes with it creates an environment in which complexity and obscurity will inevitably flourish (Grassman, 1986). My personal view is that the situation is due not so much to insidious forces as to simple lack of attention to important matters. My hope is that well-meaning scholars can and will adjust their behavior when they see that they could be much more effective than they have been. Certainly, they have a good deal to gain in respect and influence if they do.

WHY WE MODEL

There are many categories of modeling technique, including optimization, simulation, control theoretic, systems dynamics, queueing, and statistical, methods. Each of these categories is defined by conventions, terminology, standard formulations, and methods. Specific models usually fall into one of these distinct categories, although one occasionally encounters hybrid models that cut across the boundaries. All of these modeling techniques have had their capabilities extended greatly over the past few decades. With the increase in power and availability of computers, we have been able to deal with many more parameters. Perhaps the possibility of doing what we could not do before has lured us into accepting without question the view that more detail is better.

We usually do not think very much about what a model is for, since we all understand what these various techniques do. Let us say, to be general, that the basic purpose of any model is to expose the truth about some aspect of reality. However, and this is the point I wish to emphasize, exposure involves more than discovery; the truth must be understood and believed in order for it to carry any influence in the making of practical decisions. In seeking technical validity, being sure that what we think is true really is, and is not just an artifact of the model, we can easily fall into the trap of building so much into the model that the details or the structure conceal the very truth we want to expose.

Even inexperienced interpreters of models know intuitively that many things can go wrong in models: assumptions might be faulty, the data that went in might have been wrong, the computer program might have bugs, results might be misinterpreted, and so forth (Houston, 1985). Even after the creators of a model have established to their own satisfaction that the model is technically valid, few people are gullible enough to accept the creators' conclusions at face value. Particularly when the details are hidden within computational chains and loops inside a computer, most of us have learned from experience to maintain a healthy skepticism about artificially generated data. The larger and more complicated the model, the more skepticism is appropriate. Thus, we are left with the ironic dilemma that the harder we try to be correct, the less likely we are able to convince others that we are.

Most of us are acutely aware of the danger of believing in the results of a model when it is wrong. Let us call this a type I error, following the terminology used in statistical hypothesis testing. We usually mitigate or guard against such errors by stressing validation. The other kind of error, which corresponds to not believing that what a model indicates is correct when in fact it is, can be called a type II error. In this terminology, we say that in attempting to avoid type I errors, we often increase the likelihood of a type II error. Or, to say it yet another way, we seek validity at the expense of credibility.

Perhaps one reason we tend to neglect the possibility of type II errors is that credibility involves the perceptions and psychology of the beholder, whereas validity is more a property of the model itself. Another factor may be a greater fear of the dangers of a type I error: being wrong seems worse than not being believed. Nevertheless, these two types of errors are equally detrimental to the successful application of models.

Apart from issues of credibility, another consequence of excessively complex models is decreased generality. As we add details or complicate the structure, we are forced to make more and more assumptions. Although these assumptions may be entirely valid for the situation at hand, the increased specificity limits the range of applicability. If conditions change, or slight variations need to be considered, the model may no longer apply.

Yet another deficiency of complex models is the cost of developing and operating them. If the time required to collect the data necessary to run a model is excessive, or the expertise required to interpret the results is unavailable, or the time required to obtain results exceeds the time available for considering the decision, then the model cannot be of much help.

THE POWER OF SIMPLICITY

Turning from these criticisms of complex models to the advantages of simple models, I propose the following generalization. *The power of a model or of a modeling technique is a function of validity, credibility, and generality. Usually, the simplest model that expresses a valid relation will be the most powerful.* By emphasizing the power of a model as a more comprehensive measure of its utility than validity alone, I hope to encourage attention to these other aspects.

As examples of powerful models, I could cite the laws of thermodynamics, Newton's laws of force and motion, and many other familiar "elementary" equations of science. All of these are extremely simple to state but profound in their application. There are a few such laws that apply to manufacturing, such as Little's equation (see Little, in this volume), but surprisingly few such powerful relations have yet been discovered. My hope is that a deliberate effort to find them would be fruitful. This view can be expressed in a second proposition: *It is neither necessary nor desirable to build complicated models to deal with complicated situations. Indeed, we should be trying to find a point of view that makes complicated situations seem simple.*

Of course, we must be aware that simple does not mean trivial or obvious. We cannot define relations arbitrarily, make capricious assumptions, or generalize recklessly. Einstein is reputed to have said, "Things should be as simple as possible, but no simpler." Finding the adequate level of detail, the appropriate assumptions, and the elegant formulation is a matter of hard work and inspired wisdom (and perhaps a large dose of luck).

I believe that at least part of the reason that we have few simple models available to us in manufacturing is that we have not yet made a serious effort to define them. In striving to exercise our techniques to their extreme limits, and captivated by our new-found abilities to compute, we may have overlooked opportunities that would benefit us more. It may be too much to hope for laws of manufacturing that are as simple and powerful as the laws of thermodynamics, but we can certainly redirect our emphasis to developing models that are as compelling in their credibility and generality as they are impressive in technique. The goal should be to find not how much can be put in but how little will suffice to get the job done.

AN EXAMPLE

The following discussion is intended only to illustrate how a simple model can expose a general truth in a way that is practical, rigorous, and convincing. The model tracks the progression of a single product unit through stages of completion. It does not matter what the product is, what processes are involved, how long they take, or where the boundaries of the system are drawn.

A *product unit* is the thing produced. It may be a set of objects (e.g., a batch or an order), and it will generally undergo physical changes as it advances. A product unit could consist of a single workpiece, a subassembly, a batch, or a group of assemblies. A *manufacturing unit* adds value to product units while consuming time. Each of these terms is defined in such a way that the same term may apply at both atomic and aggregate levels. For example, a manufacturing unit could be a single machine, a group of machines, a department, a plant, or even an entire industry. This point is important to achieving generality in our results. When we state a result that applies to units of this degree of abstraction, the result will be applicable in many ways.

There are other kinds of units a product unit may pass through in completing its processing. For example, a *transport unit* changes the physical location of product units while consuming time, whereas a *storage unit* is passive with respect to value and location, but consumes time. A manufacturing unit may contain other manufacturing units, as well as transport units and storage units.

In this model of manufacturing, only two things happen: time advances and value is added. For any stage of the process, we can portray the change in these two dimensions. We may do this either continuously or at arbitrarily selected times. We will make no assumptions about mathematical continuity, differentiability, or monotonicity of the changes. Although time is irreversible, we could have situations in which value is lost, such as when a process is destructive.

We can now suggest a convenient visualization of a manufacturing unit on a value-time scale, from the point of view of a single product unit (see Figure 1). The effect of the manufacturing unit is defined completely by its starting time and value and its final time and value. We can therefore portray the changes in these two dimensions as a rectangle in the value-time plane. We do not necessarily have to know anything about what takes place within that rectangle; we can treat it as a "black box." Note that this picture does not consider a stream of products—just what one product unit sees in passing through an arbitrarily defined manufacturing unit.

We can string several such rectangles together to portray what a single product unit sees in passing through a sequence of manufacturing units, as

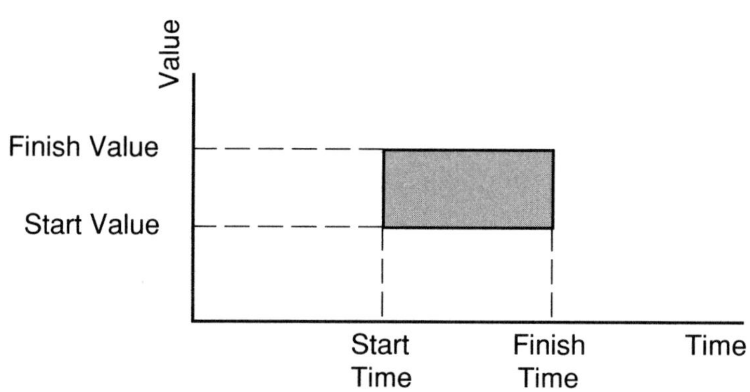

FIGURE 1 One manufacturing unit.

shown in Figure 2. The level-value lines connecting the rectangles represent the passage of time with no increase in value; hence they represent storage or transport units or possibly other forms of delay. (Incidentally, it is not necessary to assume that the value remains constant during these periods. The level lines are just special cases of rectangles, in which the height is zero. Furthermore, we could treat cases in which value declines during periods of storage, as in the case of perishable commodities, with no change in our approach.)

The different rectangles might correspond to workstations or departments in a factory, or, at another level of aggregation, they might represent different plants or warehouses. In a model of the processes involved in new product development, the rectangles might represent design, planning, or tooling. We could also refine the view to see what happens in a manufacturing unit represented by one of the rectangles, as shown in Figure 3.

We can nest manufacturing units, transport units, and storage units within manufacturing units. We can aggregate or refine to any level without altering the fundamental idea that a manufacturing unit increments value in a period of time.

As an early indication of the value of this visualization tool, let us imagine that we would like to improve the process represented in Figure 2. Where should we focus our efforts?

Define the "value improvement rate" to be the amount of value increase that a product unit receives by passing through a manufacturing unit divided by the time required. Graphically, this would correspond to the slope of a line drawn from the lower left corner of the rectangle to the upper right corner (see Figure 4).

All other factors being equal, we can improve the process within some overall manufacturing unit by either increasing the amount of value added or decreasing the time required; graphically, we would like to increase the

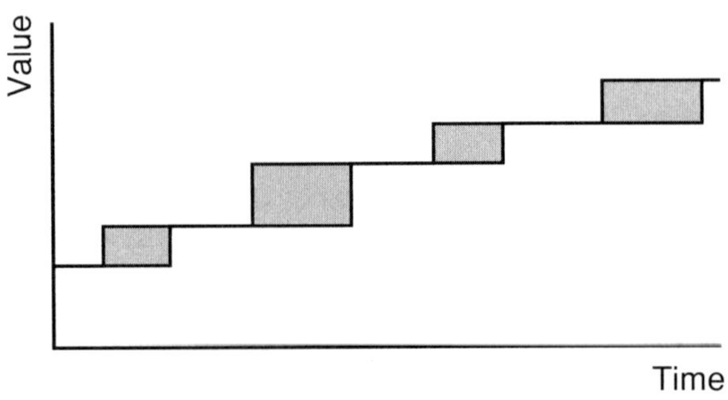

FIGURE 2 Several manufacturing units in sequence.

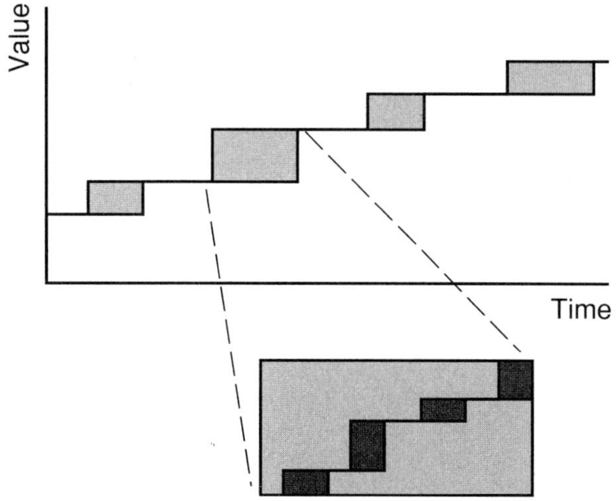

FIGURE 3 A refined view of a single manufacturing unit.

slope of the diagonal. The same interpretation applies at any level of aggregation.

A typical process improvement program will focus on the technology or management involved in just one of the manufacturing units. For example, a just-in-time production control program might be introduced in the unit represented in the second box in Figure 2—perhaps a plant or one department of a plant. However, we can easily see in Figure 5 that increasing the slope of the second box cannot alone have much of an effect upon the overall slope of the aggregate unit. In fact, if the time saved is just con-

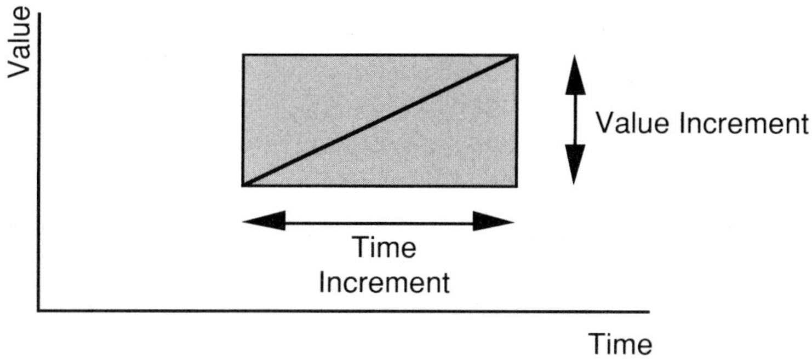

FIGURE 4 Value improvement rate is given by the slope of the diagonal.

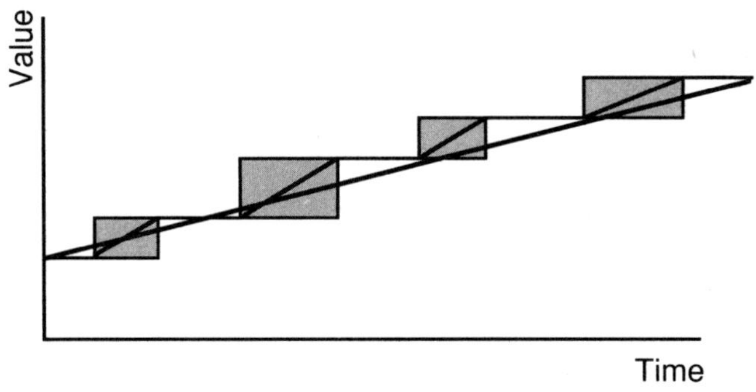

FIGURE 5 The slopes of each unit and the overall slope.

verted into an equivalent time in transport or storage, then the slope of the diagonal of the aggregate manufacturing unit would not be changed at all. This is a fairly obvious point once stated. However, the picture gives strong visual confirmation of the notion that working on one subsystem does not necessarily improve the performance of the whole system.

If Figure 2 or 5 represented a complete system, it would be apparent that any effort to improve the system should go into reducing the time lost in transit or in storage between the four subunits, since each of these has an individual slope that is better than the overall slope. These delay times between manufacturing units are frequently neglected, simply because they fall in the "hand-off" between departmental jurisdictions. In other cases, of course, one unit may stand out as the subsystem needing attention.

It may occur to you that an algebraic model could accomplish what we have shown here, with the added advantage of quantifying the parameters. You may also think of further embellishments that could be added to the model, such as costs or quality measures. If you are tempted to see what you can do with these extensions, you are experiencing the very proclivity to complicate that I have been warning against. However, you are invited to use or vary the model as you please. You may also want to consult a paper by Sullivan (1986) that proposes a similar modeling concept.

CONCLUSIONS

Some 20 years ago, John D. C. Little wrote a particularly cogent paper on the failure of operations research techniques to achieve much acceptance, for some of the same reasons expressed here (Little, 1970). Although his remarks related primarily to applications in marketing, the same comments are equally suited to manufacturing, and the passage of time has done little to change the situation.

By recognizing that the power of a model is directly proportional to its simplicity (rather than its complexity), we may be able to redirect some of our attention to new formulations. I do not expect these to be easy to generate, for it is harder to know what can safely be left out than it is to just put everything in. But there is no doubt that both the user community and the research community would be better served by models that are elegantly simple than by more elaborate constructions that go unused.

Having an arsenal of such simple models available, along with the more detailed techniques that have already been developed, would allow a more rational choice of appropriate complexity. The simple models could be used in preliminary studies, perhaps to identify key issues or to justify further investigations. More elaborate models could be used to fine tune or to explore deeper issues. It is hoped that such a spectrum of possibilities would provide greater acceptance of what the modeling community has to offer.

Improving Manufacturing Competitiveness Through Strategic Analysis

G. KEITH TURNBULL, EDEN S. FISHER,
EILEEN M. PERETIC, JOHN R. H. BLACK, ARNOLDO R. CRUZ,
and MARYALICE NEWBORN

Every business faces a dynamic environment. Customers' needs and expectations change, competitors gain strength, cost pressures increase, technological choices broaden, new markets grow. Therefore, each business must consider several questions. "What are the most significant forces at work? What impact might they have on my business? What strategic options are available to my business, and what are their implications?" The existence of options affirms that the future cannot be strictly predicted; judgment will always be essential to strategic decision making. Judgment may be enhanced, however, with insightful analysis based on facts that can be known today.

A systematic search for the facts and fundamental principles relevant to a business's strategic situation will reap a surprisingly rich harvest of information. Much of this information already exists but must be pulled together from disparate sources within the organization, including operating, marketing, research, engineering, and management people. Other important information must be drawn from outside sources, including customers and technical experts. A truly effective analysis process brings together the essential facts and specialized knowledge necessary to understand the challenges and opportunities facing a business, fosters insights that enhance decision making, and catalyzes the transition from planning to action.

CHARACTERISTICS OF STRATEGIC ANALYSIS AT ALCOA

We believe that the most consistently successful manufacturing enterprises will be those that

- Strive to anticipate systematically the important forces operating on their business and industry.
- Consider the full potential of their manufacturing systems, including the nature and implications of the changes that might be made.
- Develop the commitment and agility to capture the strategic opportunities that can be recognized through heightened awareness of external forces and internal potential.

At Alcoa we practice a strategic analysis method that reflects this belief.

Descriptions of other strategic analysis processes are widely available. What are the important distinguishing characteristics of the Alcoa process? As we see them, they are as follows:

1. Emphasis on using data for process understanding.
2. Development of forecasts for key processes, using the constraints of limits.
3. Systematic summarization of forecasts and interpretation of issues and opportunities.
4. Shared engagement in the analysis by key operating, technical, and management contributors.

Each of these points is expanded upon in the following pages. Using this approach, we are setting the strategic directions to guide a corporation that has begun its second century.

USING DATA FOR PROCESS UNDERSTANDING

Strategic analysis at Alcoa is designed to focus and enhance strategic judgment through the careful consideration of facts. Analysis participants are asked to set aside expectations associated with their prior mental models of their manufacturing system, customers, and industry, and to work to discover the implications of models true to the data that have been assembled.

A sound strategic analysis requires the compilation of a rich set of data. Data bearing on the fundamentals of any manufacturing enterprise must be included: the factors driving overall demand for its products, customer satisfaction criteria and performance against those criteria, the nature and strengths of the competition, and the efficiency and effectiveness of the manufacturing system. In addition, the analysis should also address the validity of existing mental models of the business by including objective data that supports, refutes, or refines them.

The purpose of data in strategic analysis is to contribute to a better understanding of the key processes for the business. Therefore, the data are assembled in the language of the processes themselves, for example, energy consumption in British thermal units per pound, productivity in pounds per hour, and delivery performance in days.

Process understanding for complex systems may occur at different levels. At the top level, it is valuable to have a balanced, descriptive understanding of the process, including measures of efficiencies, quantities of output, and the attributes of the output. The most profound process understanding, however, requires an examination of the underlying principles, mechanisms, and root causes.

For the manufacturing system, the following tools can provide helpful structure for identifying and organizing data that will contribute to process understanding: SIPOC (supplier-input-process-output-customer) models, flow charts, cause-and-effect diagrams, Pareto diagrams, output equations, mass balances, and energy balances. This list is illustrative rather than prescriptive; for example, a method to account for any resource believed to be strategic could be added.

The tools for structuring data collection in areas beyond manufacturing are similar, as the goal is still process understanding. Just as we try to understand the factors behind manufacturing, we try to understand the factors that drive customer satisfaction and demand.

Because of their power to communicate, graphical representations are generally used for strategic analysis data. The graphs that are used most frequently depict the measurement of a significant process feature over a long time horizon. Each such graph gains additional value when it is part of a set of graphs that make it possible to drill down through the system and approach the underlying principles. In addition to time series information, other data describing specific events, conditions, or relationships will often be useful for determining the root causes of process performance. Control charts, or other representations of process variability in the recent past, are particularly valuable. Graphs of financial measures are considered less frequently, as they do not generally promote understanding of the underlying processes.

DEVELOPMENT OF FORECASTS

The future success of a business will be influenced both by processes over which the business has little control and by those it can affect directly. For a process in the former category, we are interested in forecasting its expected performance over time. For a process in the latter category, we are interested in forecasting its potential performance, based on our understanding of "what could be" and our capacity to act. In the first instance, process understanding improves our ability to forecast opportunities that are likely to exist. In the second instance, process understanding improves our ability to identify opportunities that we could create.

The discussion that follows focuses on developing forecasts of manufacturing process potential. Manufacturing is emphasized because it is often the area in which we have the greatest ability to affect the processes and realize strategic advantage. Our approach to forecasting is also easily described in a manufacturing context, where the relationships between processes and underlying principles are relatively straightforward.

The historical time series graphs for the most significant manufacturing process features become the platform for the interactive approach we use to develop forecasts of manufacturing process potential. In a forum that brings together individuals with operating, engineering, and management skills and fundamental understanding of the key system processes, potential process opportunities are thoroughly explored against the backdrop of historical data. The exploration is driven by questions about the historical performance, theoretical limits, engineering limits, relevant benchmark information, and potential "enablers" for improving process performance (see Figure 1).

In forecasting process potential, the following questions about historical performance are important: "What explains the current situation? What has driven the rate of change in process performance? What were the factors behind the most striking characteristics of the record of our past

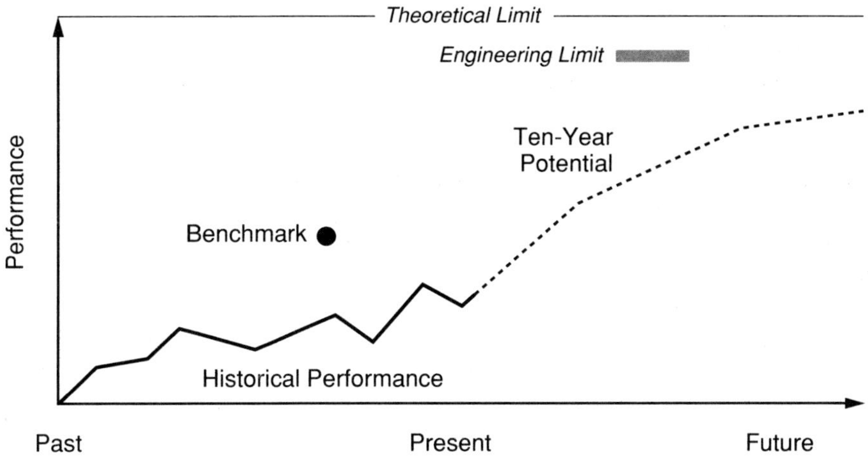

FIGURE 1 Historical and potential performance of a manufacturing process are shown in relation to its theoretical limits, engineering limits, and benchmark information.

performance?" A careful understanding of historical performance provides the basis for forecasts of future process potential.

Understanding theoretical limits provides both an outer bound for forecasts of potential process performance and a framework for clarifying the principles that govern the process. Theoretical limits are not goals or plans; they are numerical estimates of the ultimate conceivable level for a process variable, supported by the fundamental principles and reasoning that led to that numerical estimate. For some process variables, the theoretical limits will rest on laws of physics or chemistry. For other process variables, useful theoretical limits can be established using concepts of absolutes, such as the "zeros" behind just-in-time manufacturing (e.g., "zero defects," "zero breakdowns," and "zero lead time") (Fallon, 1986). The focus of this exploration is, "What are the phenomena that determine process boundaries?

Discussions of process limits may involve the identification of combinations of processes, and combinations of limits, contributing to overall process performance. Several layers of process understanding may be involved. For example, to understand the potential capacity of a furnace for heating product within our manufacturing system, both thermal capacity and physical capacity must be considered; either may limit the system. In considering the thermal capacity, contributing processes include the delivery of energy by the furnace, the effective acceptance of energy by material in the furnace, the effective use of time, and the effective conversion of materials. Each of these contributing processes can be further disaggregated, until the processes being considered are directly measurable and the operative fundamental principles can be clearly described.

Theoretical limits often introduce stretch into considerations of future process potential. Sometimes, however, considering historical performance against the theoretical limit will indicate that there is little further opportunity for improvement in a particular process feature; this is also an important strategic consideration. For example, a hundred years of technological advances in the Hall process for aluminum production have brought the parameter "current efficiency" to levels approaching what is theoretically possible.

When considered together, historical performance and theoretical limits provide general insight into the likelihood of significantly improving process performance through the investment of additional effort. In general, the "S-curve" phenomena predicts that process performance improvements in response to invested effort will reflect a rapid learning phase, followed by a period of diminishing returns as the limit is approached (Foster, 1986).

To an individual business faced with a strategic choice, however, the specific nature of the most effective "invested effort" is critical. Therefore, it is useful to consider the underlying data that describe causes for the gap between current performance and the theoretical limit. For example, what are the sources of defects or downtime? Is there benchmark information—from sister manufacturing facilities, competitors, or even an unrelated industry—that indicates that someone else has partially closed the gap? If so, how? What enablers might be employed to approach the theoretical limit?

By broadly searching for ways to move an indicator of process performance from historical levels, an engineering limit for that variable can be established. We define engineering limits as numerical estimates of the levels process variables could attain, using known technologies. When some of the known technologies under consideration would involve unusually large investment or expense, two sets of engineering limits might be established, representing different assumptions about available financial resources.

Engineering limits are not goals or plans. Although the engineering limit for a specific process performance indicator is intended as an estimate of what could actually be achieved, it may not consider possible adverse effects on other performance indicators. Developing these limits, however, does sharpen understanding around what actions might be taken to improve process performance.

The framework of historical performance, benchmarks, and theoretical and engineering limits provides a rich basis for developing 3-, 5-, or 10-year potentials for process performance. These quantitative estimates of "what could be" reflect the impact of the enabling factors that the assembled forum believes could be applied to their manufacturing system over the specified time period. In developing process potentials, participants also consider the implications of interactions among process indicators.

Overall, the development of process limits deepens the participants understanding of how key processes might perform in the future and how specific actions can influence this outcome. For each process that is examined, potential actions are considered within the context of such questions as the following: Can improvement efforts make a significant impact, or is this process nearing its theoretical limit? What would be the impact on us, and on our customers, of significantly reducing current levels of process variability? Are there alternative technologies, representing rapidly improving processes that could overtake the existing process? Can process steps be eliminated?

The most visible product of the exploration is a set of graphs that include quantitative estimates of manufacturing process potentials over the next 3, 5, or 10 years. Behind these numbers, however, are shared mental models of how key processes in the system behave and shared understanding of the issues associated with striving to approach process boundaries (see Figure 2).

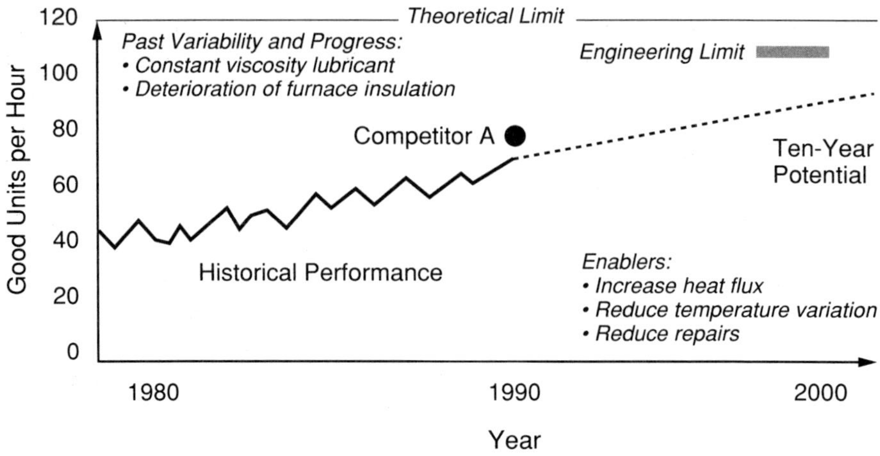

FIGURE 2 Quantitative estimates of "what could be" when appropriate enabling factors are applied to the manufacturing system.

SUMMARIZING FORECASTS AND INTERPRETING OPPORTUNITIES

Developing the forecasts of manufacturing process potentials on simple graphs of single parameters allows individuals with diverse perspectives on the same manufacturing enterprise to build a common understanding of process-specific issues. Before the strategic implications of the manufacturing process forecasts can be harvested, however, a meaningful summary of the manufacturing process analysis is essential. In addition, there must be careful consideration of the interplay among the manufacturing process potentials and forecasts of customer satisfaction, demand, and competitor performance.

We have found two types of summaries to be of value in strategic analyses: a summary of opportunities for the current manufacturing system and a summary of opportunities for the business. The manufacturing summary assumes that the current manufacturing facilities represent the essence of the production system and examines the opportunities associated with manufacturing products similar to those that are currently produced. The business summary looks more broadly at the relationships among forecasts of our manufacturing capabilities and the forecasts of customer needs, market demand, and competitor strengths.

The ultimate manufacturing summary is an estimate of the total financial opportunity associated with reaching the forecast manufacturing process potentials. This financial opportunity includes both a value for the potential cost savings at current production levels that could be realized through process improvement, and a value for the potential financial impact of increased output from the manufacturing facilities.

In the financial summary of manufacturing opportunity, the manufacturing process potentials are expressed in a language that facilitates consideration of the potential impact of manufacturing improvements. Yet, this estimate ties directly back to the individual explorations of manufacturing process parameters. Current values and forecast potential values and both historical and potential rates of change are all arrayed in a supporting table. In this way, the range and magnitude of the improvements required to capture the estimated financial opportunity is apparent.

The business summary builds on the manufacturing summary by assessing the manufacturing performance potentials against the forecasts of customer needs, market demand, and competitor performance. Consider the following questions: "What are the forces acting to change customer needs, market demand, manufacturing quality, and competitor production? How will my product offerings track with customer needs? How will my operating capabilities track with industry trends? How will changes in my production capacity compare with changes in demand? How could my position change relative to the competition?" Examining the historical and forecast rates of change in the data base that has been developed provides fact-based insights that help define the strategic situation for a business.

Because the forecasts of process potentials were established through a process that identified enablers for progress, another element of the strategic situation that can be summarized relates to the capabilities of the business. What common themes in resource needs, skills, or problem solving are evident from grouping the enablers for improved process performance? Which capabilities are the most important to develop? How would the business be different, once those capabilities were in place? This exploration of potential capabilities poises the business to consider an expanded set of customers and markets in the future.

SHARED ENGAGEMENT IN THE ANALYSIS

Much of the power of the strategic analysis process is due to the fact that the strategic issues for the business are identified by people in the business. Participation in the process cuts across a number of job levels and brings together operating, marketing, research, engineering, and management contributors. The engagement of the human system at several levels improves the analysis by drawing upon the knowledge of those closest to the processes. It also facilitates deployment of strategic decisions because many of the individuals who will be involved in changes have contributed to the process of identifying the opportunities associated with change.

Participation is not limited to people in the business, however. Because of our belief in the strategic importance of science and technology, outside experts in relevant fundamental principles and technologies should also be drawn into the analysis. Because the strategic analysis process itself, and particularly the development of forecasts using limits, benefits from experienced management, members of Alcoa's Technology Planning Division provide leadership in each analysis. Because the set of strategic issues raised by each business unit has implications for other business units and for the corporation as a whole, Alcoa's corporate Management Committee also participates in each strategic analysis.

The process steps in a strategic analysis require engagement and meaningful contributions from all of the above. Strategic analysis requires comanagers from the business and the Technology Planning Division, along with a sponsor from business management, to identify the scope of the analysis and the key participants at each step. Operating personnel and technical experts are essential to forecast and summarize manufacturing process data. Broad representation from the business engages in force field analysis, which is used to identify the business implications of relationships among manufacturing, customer needs, market demand, and competitor forecasts. The members of Alcoa's corporate Management Committee draw upon their individual areas of expertise and their insight into the corporation as a whole to provide important reviews of, and inputs to, the strategic analysis, the strategic options that the business subsequently develops to highlight the opportunities that it could pursue, and the strategic plan that is finally established. The members of the Management Committee are customers of the strategic analysis process as well as contributors, because the findings of the individual business unit efforts are used in the formulation of corporate strategy.

Although the participants have different roles, all engage in data analysis, develop process understanding and practice fact-based decision making. Therefore, strategic analysis not only enhances judgments about strategic issues but also strengthens core skills of the individuals in the organization.

ACKNOWLEDGMENTS

The strategic analysis process at Alcoa is built on a platform of several years of technology forecasting and planning experience with Alcoa businesses. The "Hall Process" example mentioned here was the work of Joseph Tribendis, who helped establish the forecasting method described in this paper. Ron McClure, Dick Wehling, and Mike Bresko also made important contributions to early stages of the development of this methodology.

Going to the Gemba

ALBERTUS D. WELLIVER

Every so often we hear about businesses making dramatic comebacks, struggling out from a quagmire of red ink and scrambling up to world-class ground. While improvement in these cases is remarkable, the threat of imminent bankruptcy is usually the sole motivating force. On the other hand, the adrenalin seems to run dry in companies that have stayed successful over an extended period of time. When businesses are riding high, how can corporate leaders instill a sense of urgency among employees to improve their work? Or expressed another way, how can the "fat cats" in corporate America continue to savor the thrill of the hunt?

To find the answers to such questions, we need to take a look at some basic quality principles. At its core, continuous improvement has to do with the way we promote and manage change. In western businesses, we usually look for big changes that reap big results in a short time, especially innovations that lead to increases along the financial bottom line. In Japan and in other Pacific Rim countries, change in small increments is encouraged, often with better, longer-lasting results that are realized over the longer term. For those who are accustomed to gradual change, small improvements become a way of life that spills over into the way employees approach their work. That means, instead of making a great technological leap, then settling back until the next great upheaval, employees constantly make small adjustments to their work processes. This ongoing change, or *kaizen*, is described by Masaaki Imai (1986) as a people-oriented approach that focuses more on the continuous process of improvement than the results of that effort.

The key to emphasizing the process is what Japanese managers call "going to the gemba." In the manufacturing environment, an example of *gemba*, defined as any place where work is being done, is the factory floor. Here, managers can find facts and data that reveal where problems lie and ultimately, where to target improvements. By visiting the gemba for a firsthand look, managers do not have to rely on traditional communication lines for information, nor do they have to wonder whether the information was culled as it rose up through organizational layers.

A descriptive model, "The Iceberg of Knowledge," addresses the process of change in relation to the typical structure within an organization. The model has three elements, two of which are knowledge and power. Figure 1 shows that the distribution of knowledge in an organization is shaped like an iceberg, while the distribution of power can be described as an inverted iceberg. Technical details and awareness of problems that disrupt or slow down productivity reside with those in the gemba who do the practical work; yet power, or the ability to make changes, is embraced by top management. These managers must realize that issues are always perking near the bottom of the organization, even when the company's balance sheet looks healthy.

As a result of the imbalance between knowledge and power, high-level managers are generally not in touch with the realities of the production line,

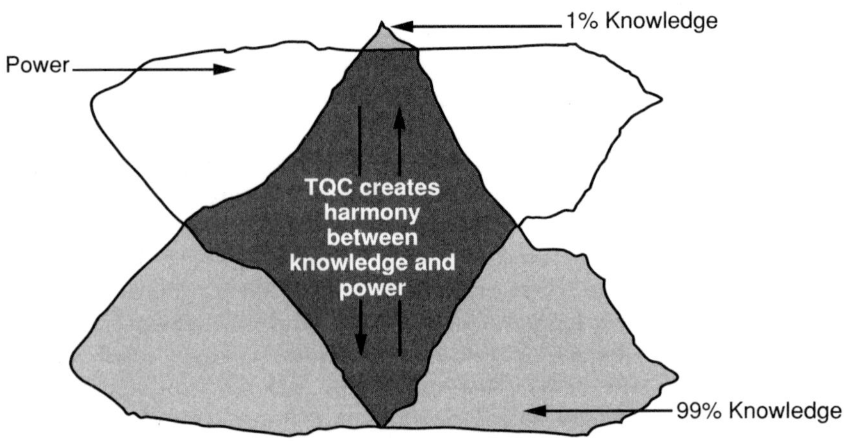

FIGURE 1 "The Iceberg of Knowledge" addresses the process of change in relation to the typical structure within an organization.

and little decision-making authority is in the hands of those who actually have the facts and data.

Referring again to Figure 1, what brings harmony to this disparity is a third element—total quality control (TQC). A basic element in any TQC effort is communication of data—specifically, statistics and information that describe a problem or establish a benchmark for improvement. Awareness of problems is what maintains the sense of urgency among managers to initiate changes that lead to improvement. But this type of knowledge is difficult to gain in large corporations. Sheer size can create organizational obstructions to the flow of information. Too often, the layers of management in an organization work as a filter, rather than a conduit, for facts and data.

If an organization is to successfully absorb change, whether it is brought on by improvements in technology or processes, a system must be in place that allows communication throughout all hierarchical levels.

Communication is initiated when senior management clearly states company goals and objectives for all to discuss and understand, then facts and data are invited directly from employees to help pinpoint where improvements are needed. Essentially, this exchange helps integrate knowledge and power.

As the flow of information increases, people with the practical knowledge will regularly help form organizational goals and objectives in a give-and-take process, which the Japanese call "catch-ball." The manager would say, "Here's what I'd like you to do." And the employee would respond by saying how he thought he could do it. They would toss ideas back and forth until they finally agreed. In America, we are more apt to play hardball. Top management says, "do it or else," and if accomplishments fall short, the employee is hit with demerits.

At Boeing, we have initiated steps to establish a communication system that will bring together knowledge and power. At one fabrication plant, the equipment operators have been placed in charge of statistical process control.

That is quite a change from the way it used to be. The operators used to be judged by how many parts they turned out, regardless of accuracy. Wasted effort accumulated quickly using this method. For example, if a part has to go through n machining actions at different stations before it is complete, the operator at one station would usually produce these parts in lots and pass them on, without inspection, to the next station. If each machining task took an hour to perform, it would be n hours before the industrial engineer, who inspects parts only on completion, discovered that a hole was drilled a fraction off the mark. If the design requires a smaller variance, this process has not only escalated waste in material and time, but contributed to human frustration.

Now that Boeing better understands the value of the gemba, the operators have charge of the process. They collect data on every part they manufacture, measuring where that part distributes within a range of variance. Because they inspect their own work, feedback is immediate. They can adjust their machine, and if it is not exact enough, they have the authority to shut it down until someone from maintenance can get the machine working up to standard.

What has happened at this factory simply follows human nature. When operators record data on a large tablet or board, it is readily visible not only to them, but to their supervisors. And the operator who uses a particular machine on the next shift can rely on that information to help set up the machine to begin his work. As the data accumulate, managers are able to identify problems and work with operators to tighten production variance gradually. Managers have entrusted their employees to think and plan their next move, not just act. This is when quality starts to improve. Moreover, when managers have access to data that come straight from the factory floor, they know where to focus their improvement efforts.

This change in approach at Boeing has resulted in a steady reduction in waste and rework. The rise in self-esteem among employees cannot be measured, but we know it is there. Employees have been given more freedom and power than ever before. They have the authority to stop the production process and call for maintenance. At the same time, they have the responsibility to make correct parts, collect data, and improve the way they work.

A second example of how Boeing is trying to merge knowledge and power occurred in late 1989. The engineers' union flatly rejected the contract offered by Boeing. Management was caught by surprise because we had been confident going into negotiations that the engineers were basically satisfied with the contract offer. How could the management be so remote from what was really going on? We simply did not have adequate facts and data.

There were no real lines of communication, so top management decided it was time to visit the gemba. We talked to engineers and their managers by the thousands, not telling them anything, but asking what was on their minds. We learned a lot. As a result, Boeing will be changing the way it handles compensation, education, career enhancement, and its performance measurement system. The next step is communicating to the entire engineering organization what we heard, and what we plan to do about it. Then, we will go back to the gemba to find out if we are on the right track. We are going to follow that up with an employee survey to check employee feedback on a broader scale. This effort is so important that top management is going to measure its own success based on those survey results. This method of benchmarking is a significant departure from previous efforts.

Boeing is slowly learning that employee motivation grows naturally out of good communication and a balance of knowledge and power. Our biggest challenge is to instill this philosophy and approach throughout the company. Managers are accustomed to having more control over the processes, but they do not realize they will have much more control over the quality of their products or services if they let go of some of the decision making. Going to the gemba has taught us that letting go is easier than most people dare to think.

We are also learning that managers must look for problems with vigilance, gaining an understanding of the real issues and problems by poring over facts and data. It is a tough job. But once the hidden problems are revealed, managers start to realize that the appearance of a smoothly operating organization can be deceptive. Once they set up a process to find out about concerns and problems below the tip of the iceberg, the gemba becomes a continuous source of information and inspiration for change. It is this process that keeps a company on the urgent edge of continuous improvement.

Jazz: A Metaphor for High-Performance Teams

RICHARD C. WILSON

In this paper I argue that jazz can provide an enlightening metaphor for participative manufacturing management, especially to dramatize some characteristics of high-performance work teams. Continuous improvement of manufacturing effectiveness often depends on the participation and involvement of all employees. Terms such as quality circles, concurrent engineering, quality-of-work-life programs, employee empowerment, and job enrichment are used to characterize participative approaches to management. But each term evokes an imagery that may be highly dependent on a person's background and experience. Because so many of us share an awareness and enjoyment of music, we should find it helpful to use the insights that a special form, jazz, offers as a metaphor for employee involvement and empowerment. The growing literature about jazz provides considerable insight about the organizational and social dynamics of the jazz world through anecdotes about the behavior of its performers and leaders. Does this organizational research on the behavioral aspects of the jazz milieu have wider implications for the practice of employee involvement groups in other occupations? Can statements by jazz artists stimulate you to think about the nature of production jobs with a fresh viewpoint? I believe they can. In developing this argument, I admit to sharing the view of trumpeter Wynton Marsalis, who says (Helland, 1990): "Jazz music really teaches you what it is to live in a democracy. The whole negotiation of the rights of individuals with responsibility to the group." How is this view of jazz helpful in thinking about groups in manufacturing?

SOLOING AND SMALL JAZZ GROUPS

Consider the following statements by jazz artists, and their applicability to the behavior of manufacturing teams:

Dave Holland, a 44-year-old jazz bass player says (Mandel, 1989): "I've always been attracted to jazz's group context. I admire how the soloist works with the rhythm section, how the bass player interacts with the drummer. To me, the music is group music. I want any group I put together to function on that level, where everybody feels they have a place, that they can be themselves, that they can stretch their imaginations and their creative aspirations as far as they are able. So I've always encouraged as much involvement from the musicians as possible. My thing is to create a setting—and I learned this from Miles [Davis]. During the time I played with him he would create the environment for the music, then let the musicians deal with it."

Jazz drummer and leader Art Blakey says (Rosenthal, 1986): "I try to play in the rhythm section to make the soloist play, make him feel like playing. The rhythm section can make the soloist play over his top, play things he never dreamed he could play, if you get behind him. You can't have a battle up there and see how much you can play, because if you make too much noise behind him, he can't concentrate on what he wants to play. . . . You got to get out there and push him. When I'm playing for Dizzy [Gillespie] I play one way, if I'm playing with Miles I play another way."

The jazz drummer Shelley Manne gave an interviewer his definition of jazz musicians (Crow, 1990): "We never play anything the same way once."

Jim Hall the guitarist says (Balliett, 1986): "Accompanying is hearing the whole texture from top to bottom of the music around you and then fitting yourself into the right place. . . . What you're trying to do is swing, and swinging is a question of camaraderie. You could be playing stiffly, but if everybody is playing that way the group will swing. But if one person is out of sync, is dragging, it feels like somebody is hanging onto your coattails."

These quotations from drummers and bass and guitar players identify musicians' attitudes in high-performance jazz groups. In small group improvisation (fewer than eight players), the players must share a commitment to excellence demonstrated through the creativity and imagination of their improvisations. The group conditions must free the players to establish the group cohesion and interdependence of their own contributions. To achieve group excellence, each of the players must be highly skilled on his instrument and play in styles that are mutually compatible. Players who respect one another's playing are more likely to share ideas and coach each other in group playing. Communication among the players is essential during performance, as they listen to and instantly respond to each other's improvisation ideas. During performance, the effectiveness of the group is determined at the lowest level of the organization, the players themselves. They intimately share instant information about their performance, have the power to determine and modify its direction, share full knowledge of the performance technology, and immediately share the rewards of the audience response. Do these sound like the criteria for the ideal work team? Bastien and Hostager (1988) report a study of a four-person jazz group and argue that the crucial factors of shared information, communication, and attention in jazz performance have implications for the study and management of organizational innovation in other contexts such as corporate acquisitions and new industry development.

Similar factors have been identified as critical to the performance of work teams in many settings other than jazz (Buchholz and Roth, 1987; Lawler, 1986). Thus, the concept of group creativity in a jazz performance may provoke some new ideas for organizing production work. For example, the moments of creative opportunity for a jazz musician are a small but highly motivating fraction of his total professional life. Hours of uncompensated practice are required to achieve those creative moments. Would most employees be similarly motivated by the opportunity for occasional breaks from routine work to engage in a creative job experience?

The value of the small jazz group as a metaphor for a manufacturing work group is certainly weakened by a number of disparities. Most obviously, the jazz group performs for itself or for an audience, whereas the manufacturing group produces a product for an often unseen customer. For the jazz musician, soloing involves a high risk of public exposure of mistakes, and of immediate criticism. The effectiveness of a performance is instantly assessed by other members of the group, unlike manufacturing, where performance evaluation may require analysis of data several hours afterward by independent auditors. Hence gratification for the jazz artist can be immediate and satisfying (Hackman, 1990). Furthermore, the jazz musician generally feels his work is meaningful, that performance is an expression of life and an intense emotional experience. Eisenberg (1990) calls this phenomenon a "jamming" experience and suggests that similar transcendental experiences can occur in other activities, including manufacturing groups. Unlike many manufacturing jobs, however, the price of admission for the jazz musician is a lifetime apprenticeship for a highly insecure career poorly understood by the ultimate consumer. The consequences may be a high level of psychological anxiety and stress, sometimes manifested by individual health problems (Wills and Cooper, 1988).

THE LARGER JAZZ ENSEMBLE

The amount of player self-determination in jazz varies from the complete freedom of "free" jazz improvisation, to the partially structured collaborations of small groups such as the Modern Jazz Quartet or the Art Blakey and Miles Davis groups, to the limited freedom exhibited by the heavily arranged Stan Kenton and later Count Basie bands. As the number of musicians in a jazz ensemble increases, collective improvisation becomes increasingly hard to execute. Listening to all other players becomes difficult, and without some organizing structure, improvisations tend to conflict and become muddy. Perhaps this is related to the "magical number seven, plus or minus two" (Miller, 1956). Jazz ensembles of more than seven players have rarely attempted collective improvisation. Instead of relying on chord changes to a tune for the skeleton of group improvisation, larger jazz ensembles use written arrangements that provide planned opportunities for improvised solos. Thus the arranger becomes an important determinant of group performance. His arrangement, not the individual player, determines the style of the ensemble. The "big band," comprising 12 or more players, removes much of the self-determination from the individual player, limiting his creative contribution to the musicality of his section performance and to his occasional solo opportunities. The success of the big band is dependent more on the distinctive "sound" of the band, rather than the distinctive creativity of the soloists. Those big bands that survived through several eras were distinguished by a single leader (e.g., Duke Ellington, Benny Goodman, Count Basie, Woody Herman, and Glenn Miller), who established an identifiable sound for the group and dominated the selection of players and the performance style of arrangements perpetuating that sound. For example, the Basie band of the 1930s grew out of a small ensemble in Kansas City that featured "head" arrangements (ensemble arrangements improvised collectively by the band in performance) invented by the many outstanding improvising soloists in the band. In time, these soloists left the band and were replaced by others. Basie relied increasingly on arrangers to provide a continuity of style. Eventually, the arrangements dictated the distinguishing sound of the band, and personnel were chosen for their ability to play and solo in a style compatible with the Basie sound. Thus, the decision-making process was removed from the collective level of the ensemble players to a specific individual, the leader.

A similar tendency toward specialization of function, formal organizational structure, written systems and procedures, and a planned communication mechanism is found in larger manufacturing organizations. The jazz literature, however, clearly illustrates that the style of personal leadership may be nevertheless a significant determinant of the performance in even these larger organizations.

Anecdotes about the behavior of band leaders are many, and mirror the realities of interpersonal relationships in other organizations. A band leader generally retains complete power to hire and fire his musicians, and within boundaries established by the American Federation of Musicians, to set their pay levels and the conditions of their travel. His relationship with fellow musicians in the band ranges from tolerant avoidance to lifelong friendship. For example, the Bunny Berigan band was said by trombonist Ray Conniff (Shapiro and Hentoff, 1955) to be "a tight little band, just like a family of bad boys with Bunny the worst of them all. We were all friends. In fact, Bunny wouldn't hire anybody he didn't like." Truly cooperative bands were unusual (the Casa Loma Band was a notable example). Depending on the leader's attitude, band members could have anywhere from no role to a significant role in recommending new musicians, allocating parts within a section, writing arrangements for the band, and choosing numbers, soloists, and directing the band during performance. Because of the intensity of continuous travel together, quirky behavior exacerbated the relations among the members, sometimes in humorous ways and sometimes in disagreeable ways.

Trombonist Grover Mitchell describes Duke Ellington's band when he joined it (Crow, 1990): "The first night or two everybody had gotten on the bandstand and really roared. But the next two, three, or four nights maybe there would be five or six of us on the bandstand, and eight or ten guys walking around out in the audience talking to people, or at the bar. One night we were on the bandstand and a waiter came up and told Jimmy Hamilton that his steak was ready. He stepped off the bandstand and started cuttin' into a steak. Later I say to Duke, 'Man, how can you put up with this?' And he told me, 'Look, let me tell you something. I live for the nights that this band is great. I don't worry about the nights like what you're worrying about. If you pay attention to these people they will drive you crazy. They're not going to drive me crazy.'"

Bill Hughes, long-time trombonist with Basie, reports (Hughes, 1989) that he was asked to fill an empty chair in the Basie band. After a week or so, he was still uncertain about his prospects with the band. He asked Basie if he could expect to stay with the band. To which Basie replied: "You're still here, aren't you?" Hughes then asked how much he would be paid. Basie: "Don't worry; we'll take care of you."

Duke Ellington solved an absentee problem in Chicago (Crow, 1990, p. 252) when Sam Woodyard failed to show up for a week at the Blue Note. Duke hired a local drummer to replace him. When Sam finally came to work, Duke had him set up alongside the other drummer and they both played. Sam messed around with the time just enough to cause the local drummer to resign, and Duke had the band back the way he had originally wanted it.

Stories about Benny Goodman are legion. Pianist Jess Stacey said (Crow, 1990, p. 260): "With Benny, perfection was just around the corner. He was hell on intonation, too. Between each set he had me pounding A's on the piano so the saxes and trumpets could be perfectly in tune. When I went with the Bob Crosby band I had the habit of pounding A's between sets. Bob looked at me and said, 'If you keep pounding that A, I'm going to give you your five years' notice.'"

Woody Herman was once asked if the continuing turnover of personnel in his band bothered him. He said the first time some of his star musicians left the band, he was devastated. He thought he would never again have such great players. In time, as personnel changes became a continuing fact of his music world, he said he began to look forward to the newcomers' new ideas and the contributions they made to the ongoing Herd.

In contrast to small jazz groups, the leader of a big band has a strong individual role in establishing the style and the expectations about the quality of performance. He relies on written communications (arrangements) to provide the structure of the performance relationships and to indicate where individuals can contribute their own creativity through solos. Nevertheless, these anecdotes suggest how widely the leadership styles of the band leaders may differ. Although the style and discipline of the band may therefore be a reflection of the leader's personality, there seems to be no obvious correlation between leadership style and commercial success. I am aware of only two serious organizational studies of jazz orchestras (Bougon et al., 1977; Voyer and Faulkner, 1989), and neither deals with professional bands whose livelihood depends on their public acceptance. In both studies, the factors affecting organizational effectiveness and process are specified by the musicians and therefore are specific to jazz orchestras only. Analogous studies of manufacturing organizations are unknown to me. It is tempting to suggest that the big band provides a better metaphor of current manufacturing organizations than the looser structure of the small jazz groups. However, more careful comparative studies seem necessary to support such speculation. Perhaps as Lawler (1986, p. 210) suggests, the most important asset leaders of jazz orchestras share with leaders in manufacturing is their long-term vision for the organization itself.

Jazz as a metaphor for even larger multiechelon manufacturing organizations may be stretching beyond the scope of plausibility. With some reservation, then, I close with an anecdote about the Savoy Sultan Jump band. As the house band at the Savoy Ballroom, the band alternated sets with the visiting band. It was common practice for the Sultans to open their set by playing along with the last chorus of the closing number by the visiting band, and to continue playing through several choruses on their own without losing a beat. Not only was this a graphic demonstration of their musicianship, but, to the chagrin of the visiting band, they often swung much harder to boot! Can you visualize an analogous performance by workers during a factory shift change?

ACKNOWLEDGMENT

Karl E. Weick, Rensis Likert Professor of Organizational Behavior and Psychology, The University of Michigan School of Business Administration was most helpful in the preparation of this paper.

Consolidated Bibliography

Abernathy, W. J., and K. Wayne. 1974. Limits of the learning curve. Harvard Business Review September-October:109-119.

Adler, P. S., and K. B. Clark, 1991. Behind the learning curve: A Sketch of the learning process. Management Science 37(3):267-281.

Allaire, Y., and M. Firsirotu. 1984. Theories of organizational culture. Organization Studies 5:193-226.

Allan, G. B. 1975. Note on the Use of Experience Curves in Competitive Decision Making in Harvard Business School Note 9-175-174.

Allen, T. J. 1977. Managing the Flow of Technology. Cambridge, Mass.: MIT Press.

American Supplier Institute. 1989. Seventh symposium on Taguchi methods. Dearborn, Mich.: ASI Press.

Amstead, B. J., P. F. Ostwald, and M. L. Begeman. 1977. Manufacturing Processes, 7th ed. New York: John Wiley & Sons. Cited in Hayes, R. H., and S. C. Wheelwright. 1984. Restoring Our Competitive Edge: Competing Through Manufacturing. New York: John Wiley & Sons.

Argote, L., and D. Epple. 1990. Learning curves in manufacturing. Science 247:920-924.

Argyris, C. 1991. Teaching Smart People How to Learn. Harvard Business Review May-June.

Argyris, C., and D. A. Schon. 1978. Organizational Learning: A Theory of Action Perspective. New York: Addison-Wesley.

Baker, K. R. 1974. Introduction to Sequencing and Scheduling. New York: Wiley & Sons.

Balliett, W. 1986. American Musicians. New York: Oxford University Press.

Bastien, D. T., and T. J. Hostager. 1988. Jazz as a process of organizational innovation. Communication Research 15:5.

Bate, P. 1984. The impact of organizational culture on approaches to organizational problem solving. Organizational Studies 5:43-66.

Bateson, G. 1972. Steps to an Economy of Mind. New York: Ballantine Books.

Baudin, M. 1990. Manufacturing Systems Analysis. Englewood Cliffs, N.J.: Yourdon Press.

Bechte, W. 1988. Theory and practice of load-oriented manufacturing control. International Journal of Production Research 26:375-395.

Blackstone, J. H. 1989. Capacity Management. Carlsbad, Calif.: Southwestern Press.

Blackstone, J. H., Phillips, D. T., and G. L. Hogg. 1982. A state-of-the-art survey of dispatching rules for manufacturing job shop operations. International Journal of Production Research

Blake, R. R., and J. S. Mouton. 1981. Productivity: The Human Side. New York: AMACOM.

Bloch, E. 1985. Manufacturing technologies. The Bridge 15(Fall):10-15.

Bloch, E. 1991. Toward a U.S. Technology Strategy: Enhancing Manufacturing Competitiveness. Discussion paper no. 1, The Manufacturing Forum. Washington, D.C.: National Academy Press.

Bougon, M., K. Weick, and D. Binkhorst. 1977. Cognition in organizations: An analysis of the Utrecht Jazz Orchestra. Administrative Science Quarterly 22:610.

Brech, E. F. L. 1958. Organization, The Framework of Management. London: Longmans, Green and Co.

Bridges, W. 1988. Surviving Corporate Transition. New York: Doubleday and Company.

Buchholz, S., and T. Roth. 1987. Creating the High Performance Team. New York: John Wiley & Sons, Inc.

Buck, J. R., J. M. A. Tanchoco, and A. L. Sweet. 1976. Parameter estimation methods for discrete exponential learning curves. AIIE Transactions 8(2):184-194.

Camp, R. C. 1989. Benchmarking: The Search for Industry Best Practices that Lead to Superior Performance. Milwaukee, Wisc.: Quality Press.

Chew, W. B., T. F. Bresnahan, and K. B. Clark. 1990. Measurement, coordination, and learning in a multiplant network. Pp. 129-162 in Measures for Manufacturing Excellence. R. S. Kaplan, ed. Boston, Mass.: Harvard Business School Press.

Clark, G. M., and D. H. Withers. 1989. Architecture for an integrated simulation/CIM system. Pp. 942-948 in the Proceedings of the 1989 Winter Simulation Conference, E. A. MacNair, K. J. Musselman, and P. Heidelberg, eds. Washington D.C.

Clark, K. B., and T. Fujimoto. 1989a. Overlapping problem solving in product development. Pp. 127-152 in Managing International Manufacturing, K. Ferdows, ed. North Holland: Elsevier Science Publishers B.V.

Clark, K. B., and T. Fujimoto. 1989b. Reducing the time to market: The case of the world auto industry. Design Management Journal 1:49-57.

Clark, K. B., and T. Fujimoto. 1991. Product Development Performance. Cambridge, Mass.: Harvard Business School Press.

Cohen, S. S., and J. Zysman. 1988. Manufacturing innovation and american industrial competitiveness. Science 239:1110-1115.

Compton, W. D., ed. 1988. Design and Analysis of Integrated Manufacturing Systems. Washington D.C.: National Academy Press.

Conway, R. W., W. L. Maxwell, and L. W. Miller. 1967. Theory of Scheduling. Reading, Mass.: Addison-Wesley.

Cook, H. E. 1991. On competitive manufacturing enterprises II: S-model paradigms. Manufacturing Review June.

Cook, H. E., and R. E. DeVor. 1991. On competitive manufacturing enterprises I: The S-model and the theory of quality. Manufacturing Review June.

Cooper, K. G. 1980. Naval ship production: A claim settled and a framework built. Interfaces 10(6):20-36.

Cooper, R., and R. S. Kaplan. 1991. Profit priorities from activity-based costing. Harvard Business Review May-June.

Copley, F. B. 1923. Frederick W. Taylor: Father of Scientific Management, 2 Vols. New York: Harper & Row.

Crow, B. 1990. Jazz Anecdotes. New York: Oxford University Press.

Davis, S. 1987. Future Perfect. Reading, Mass.: Addison-Wesley.

Dertouzos, M. L., R. K. Lester, R. M. Solow, and the MIT Commission on Industrial Productivity. 1989. Made In America: Regaining the Productive Edge. Cambridge, Mass.: The MIT Press.

Development Project Study. Manufacturing Vision Group

Dixon, J. R., A. J. Nanni, and T. E. Vollmann. 1990. The New Performance Challenge: Measuring Operations for World-Class Competition. Homewood, Illinois: Dow Jones-Irwin.

Drucker, P. F. 1985. Innovation and Entrepeneurship: Practice and Principles. New York: Harper & Row.

Drucker, P. F. 1988a. The coming of the new organization. Harvard Business Review January-February.

Drucker, P. F. 1988b. Management and the world's work. Harvard Business Review September-October.

Drucker, P. F. 1990. The emerging theory of manufacturing. Harvard Business Review May-June.

Ealey, L. A. 1988. Quality by Design. Dearborn, Mich.: ASI Press.

Eccles, R. G. 1991. The performance measurement manifesto. Harvard Business Review January-February.

Edmondson, H. E., and S. C. Wheelwright. 1989. Outstanding manufacturing in the coming decade. California Management Review, Vol. 31, Number 4.

Eisenberg, E. 1990. Jamming, transcendence through organizing. Communication Research 17:139.

Evans, J. R., D. R. Anderson, D. J. Sweeney, and T. A. Williams. 1984. Applied Production and Operations Management. St. Paul, Minn.: West Publishing Company.

Evans, J. R., and W. M. Lindsay. 1989. The Management and Control of Quality. St. Paul, Minn.: West Publishing Company.

Fallon, M. 1986. The seven zeroes of JIT. Managing Automation May:76.

Foster, R. N. 1982. A call for vision in managing technology. Business Week, May 24.

Foster, R. 1986. Innovation, the Attacker's Advantage. Summit Books.

French, S. 1982. Sequencing and Scheduling. Chichester, United Kingdom: Ellis-Horwood.

Galbraith, J. R. 1977. Organization Design. Reading, Mass.: Addison-Wesley.

Garvin, D. A. 1984. What does 'product quality' really mean? Sloan Management Review Fall:25-43.

Gibson, J. E. 1981. Management style and structure. Chapter in Managing Research and Development. New York: Wiley.

Gomory, R. E., and R. Schmitt. 1988. Science and product. Science 240(26 May):1131.

Grassman, W. K. 1986. Is the fact that the emperor wears no clothes a subject worthy of publication? Interfaces 16(2):43-51.

Greene, J., ed. 1987. Production and Inventory Control Handbook, 2nd ed. New York: McGraw-Hill.

Haas, E. A. 1987. Breakthrough manufacturing. Harvard Business Review March-April.

Hackman, J. R., ed. 1990. Groups That Work (and Those That Don't). San Francisco: Jossey-Bass Publishers.

Harrington, J. 1984. Understanding the Manufacturing Process. New York: Marcel Dekker, Inc.

Harrington, J., Jr. 1973. Computer-Integrated Manufacturing. Melbourne, Fla.: Robert E. Krieger Publishing Company.

Hatvany, J., 1983. The efficient use of deficient knowledge. Annals of CIRP 32(1):423-425.

Hauser, J. R., and D. Clausing. 1988. The house of quality. Harvard Business Review May-June:66-73.

Hax, A. C., and N. S. Majluf. 1984. Strategic Management. Englewood Cliffs, N.J.: Prentice-Hall.

Hayes, R. H., and R. Jaikumar. 1988. Manufacturing's crisis: New technologies, obsolete organizations. Harvard Business Review September-October:77-84.

Hayes, R. H., and S. C. Wheelwright. 1984. Restoring Our Competitive Edge: Competing Through Manufacturing. New York: John Wiley & Sons.

Hayes, R. H., S. C. Wheelwright, and K. B. Clark. 1988. Dynamic Manufacturing: Creating the Learning Organization. New York: The Free Press.

Helland, D. 1990. Wynton, prophet in standard time. Downbeat 57(9):16-19.

Henderson, B., and F. K. Levy. 1965. Adaptation in the production process. Management Science 11(6):B136-B154.

Hiltz, S. R., and M. Turoff. 1978. The Network Nation. Reading, Mass.: Addison-Wesley.

Hopp, W. J., M. L. Spearman, and D. L. Woodruff. 1990. Practical strategies for lead time reduction. Manufacturing Review 3(2):78-84.

Horvath, M. 1988. Manufacturing engineering: The birth and growth of a new science. Robotics & Computer-Integrated Manufacturing 4(1/2).

Hounshell, D. A. 1984. From the American System to Mass Production 1800-1932:

The Development of Manufacturing Technology in the United States. Baltimore: The Johns Hopkins University Press.

House, C. H., and R. L. Price. 1991. The Return Map: Tracking Product Teams. Harvard Business Review January-February:92-100.

Houston, T. R. 1985. Why models go wrong. Byte 10(10):151-164.

Howard, W. G., Jr., and B. R. Guile, eds. 1992. Profiting From Innovation. New York: The Free Press.

Hughes, B. 1989. Comments at International Symposium of Jazz Educators. Workshop. San Diego, Calif.

Imai, M. 1986. Kaizen (Ky'zen): The Key to Japan's Competitive Success. New York: McGraw-Hill Publishing Company.

International Business Machines. 1987. Computer Aided Manufacturing, An IBM Perspective. International Business Machines.

International Business Machines. 1989. Computer Aided Manufacturing, The CIM Enterprise. International Business Machines.

Jaikumar, R. 1986. Postindustrial manufacturing. Harvard Business Review November-December:69–76.

Johnson, H. T., and R. S. Kaplan. 1987. Relevance Lost: The Rise and Fall of Management Accounting. Boston, Mass.: Harvard Business School Press.

Kanet, J. J., and H. H. Adelsberger. 1987. Expert systems in production scheduling. European Journal of Operational Research 29:51-59.

Kantrow, A., et al. 1983. Survival Strategies for American Industry. New York: Wiley.

Kaplan, R. S., ed. 1990. Measures for Manufacturing Excellence. Boston, Mass.: Harvard Business School Press.

Kelly, M. R. and H. Brooks. 1991. External learning opportunities and the diffusion of process innovations to small firms. Technological Forecasting and Social Change 39:103-125.

Koska, D. K., and J. D. Romano. 1988. Countdown to the Future: The Manufacturing Engineer in the 21st Century. Society of Manufacturing Engineers.

Lawler, E. E. III. 1986. High Involvement Management. San Francisco: Jossey-Bass Publishers.

Leong, G. K., D. L. Snyder, and P. T. Ward. 1990. Research in the process and content of manufacturing strategy. Omega 18:109-122.

Little, J. D. C. 1970. Models and managers: The concept of a decision calculus. Management Science 16(8):B466-485.

Loucks, V. R., Jr. 1990. Measuring success across the '90s. Review 1:15-18.

Malcolm Baldrige National Quality Award. 1991. U.S. Department of Commerce, National Institute of Standards and Technology.

Malone, T. W. 1988. What is coordination theory. Working Paper 2051-88. Sloan School of Management, Massachusetts Institute of Technology, Cambridge, Mass.

Mandel, H. 1989. Dave Holland, creative collaborator. Downbeat 56(1):20-23.

March, J. G., and H. A. Simon. 1958. Organizations. New York: John Wiley and Sons, Inc.

Maxwell, W. L., Muckstadt, J. A., Thomas, L. J., and J. VanderEecken. 1983. A modeling framework for planning and control of production in discrete parts manufacturing and assembly systems. Interfaces 13:92-104.

Merchant, M. E. 1961. The manufacturing-system concept in production engineering research. Annals of CIRP 2(1):77-83.

Merchant, M. E. 1988. The precepts and sciences of manufacturing. Robotics & Computer-Integrated Manufacturing 4(1/2):1-6.

Miller, G. A. 1956. The magical number seven, plus or minus two. Psychological Review 63:81.

Mintzberg, H. 1978. The Structuring of Organizations. A Synthesis of the Research. Englewood Cliffs, N.J.: Prentice-Hall, Inc.

Mize, J. H., and T. G. Beaumariage. 1988. A nation at risk: Our eroding skill base in manufacturing systems. Pp. 42-51 in The Challenge to Manufacturing: A Proposal for a National Forum. Washington, D.C.: National Academy of Engineering.

Moody, P. E., ed. 1990. Strategic Manufacturing. Homewood, Ill.: Dow Jones-Irwin.

Nadler, G., and G. H. Robinson. 1983. Design of the automated factory: More than robots. Annals of the American Academy of Political and Social Science 470(November).

National Academy of Engineering. 1985. Education for the Manufacturing World of the Future. Washington, D.C.: National Academy Press.

National Academy of Engineering. 1988. The Technological Dimensions of International Competitiveness. Washington, D.C.:National Academy Press.

National Academy of Engineering. 1991. National Interests in an Age of Global Technology. Washington, D.C.: National Academy Press.

National Center for Manufacturing Sciences. 1990. Competing in World-class Manufacturing: America's 21st Century Challenge. Homewood, Ill.: Business One Irwin.

National Institute of Standards and Technology. 1991. Application Guidelines: Malcolm Baldrige National Quality Award. U.S. Department of Commerce.

National Research Council. 1986. Toward a New Era in U.S. Manufacturing: The Need for a National Vision. Washington, D.C.: National Academy Press.

National Research Council. 1990. The Internationalization of U.S. Manufacturing: Causes and Consequences. Washington, D.C.:National Academy Press.

National Research Council. 1991. The Competitive Edge: Research Priorities for U.S. Manufacturing. Committee on Analysis of Research Directions and Needs in U.S. Manufacturing. Washington, D.C.: National Academy Press.

National Research Council. 1991. Improving Engineering Design. Washington, D.C.:National Academy Press.

Nevins, J. L., and D. E. Whitney. 1989. Concurrent Design of Products and Processes. New York: McGraw-Hill.

Nonaka, I. 1988. Toward middle-up-down management: Accelerating information creation. Sloan Management Review.

Nonaka, I. 1989. Creating organizational order out of chaos: Self-renewal in Japanese firms. California Management Review.

Orlicky, J. A. 1975. Material Requirements Planning. New York: McGraw-Hill.

Passmore, W. A. 1988. Designing Effective Organizations. New York: John Wiley & Sons.

Pegals, C. C. 1969. On startup or learning curves: An expanded view. AIIE Transactions 1(3):216-222.

Peters, T. 1987. Thriving on Chaos: Handbook for a Management Revolution. New York: Alfred A. Knopf.

Phadke, M. S. 1989. Quality Engineering Using Robust Design. New York: Prentice Hall.

Pirsig, R. M. 1974. Zen and the Art of Motorcycle Maintenance: An Inquiry Into Values. Toronto: Bantam Books.

Prahalad, C. K., and G. Hamel. 1990. The core competence of the corporation. Harvard Buiness School May-June.

Pritsker, A. A. B. 1986a. Introduction to Simulation and SLAM II. New York: Wiley & Sons; West Lafayette, Ind.: Systems Publishing Corporation.

Pritsker, A. A. B. 1986b. Model evolution: A rotary index table case history. Pp. 703-707 in Proceedings of the 1986 Winter Simulation Conference, J. R. Wilson, J. O. Henriksen, and S. D. Roberts, eds. Washington, D.C.

Pritsker, A. A. B. 1990. Papers, Experiences, Perspectives. West Lafayette, Ind.: Systems Publishing Corporation.

Pritsker, A. A. B., C. E. Sigal, and R. D. J. Hammesfahr. 1989. SLAM II Network Models for Decision Support. Englewood Cliffs, N.J.: Prentice Hall.

Pritsker Corporation. 1989. FACTOR Implementation Guide, Version 4.0. West Lafayette, Ind.: Pritsker Corporation.

Pritsker Corporation. 1990. SLAMSYSTEM User's Guide. West Lafayette, Ind.: Pritsker Corporation.

Quinn, J. B. 1983. U.S. Leadership in Manufacturing. National Academy of Engineering 18th Annual Meeting Proceedings, November 4, 1982. Washington, D.C.: National Academy Press.

Reich, R. B. 1991. The Work of Nations: Preparing Ourselves for 21st-Century Capitalism. New York: Alfred A. Knopf.

Reinertsen, D. G. 1983. Whodunit? The search for the new-product killers. Electronic Business 9(8):62-64.

Report of the National Critical Technologies Panel. 1991. Washington, D.C.: Superintendent of Documents, U.S. Government Printing Office.

Rosenthal, D. 1986. Conversation with Art Blakey. The Black Perspective in Music 14:282.

Schein, E. H. 1969. Process Consultation: Its Role in Organizational Development. Reading, Mass.: Addison-Wesley.

Schneiderman, A. M. 1988. Setting quality goals. Quality Progress April:51-57.

Schonberger, R. J. 1986. World Class Manufacturing: The Lessons of Simplicity Applied. New York: The Free Press.

Schonberger, R. J. 1987. World Class Manufacturing Casebook: Implementing JIT and TQC. New York: The Free Press.

Senge, P. M. 1990. The Fifth Discipline. New York: Doubleday/Currency.

Shapiro, N., and N. Hentoff. 1955. Hear Me Talkin' to Ya. New York: Holt, Rinehart, & Winston Inc.

Shingo, S. 1985. A Revolution in Manufacturing: The SMED System. Cambridge, Mass.: Productivity Press.

Shingo, S. 1988. Non-Stock Production: The Shingo System for Continuous Improvement. Cambridge, Mass.: Productivity Press.

Shingo, S. 1989. A Study of the Toyota Production System from an Industrial Engineering Viewpoint. Cambridge, Mass.: Productivity Press.

Simon, H. A. 1957. Administrative Behavior. 2nd ed. New York: The Free Press.

Simon, H. A. 1976. Administrative Behavior. New York: The Free Press.

Simon, H. A. 1990. Prediction and prescription in systems modeling. Operations Research 38:7-14.

Society of Manufacturing Engineers. 1984. Group Technology at Work. N. L. Hyer, ed. Dearborn, Mich.: Society of Manufacturing Engineers.

Sohlenius, G., 1984. Scientific and Structural Base of Manufacturing. Robotics & Computer-Integrated Manufacturing 1(3/4):389–396.

Stalk, G., Jr., and T. M. Hout. 1990. Competing Against Time. New York: The Free Press.

Stata, R. 1989. Organizational learning—The key to management innovation. Sloan Management Review Spring.

Stevens, C. H. 1989. Cotechnology for the global 90s. Management in the 1990s Working Paper 89-074. Sloan School of Management, Massachusetts Institute of Technology, Cambridge, Mass.

Stewart, T. A. 1991. The new American century: Where we stand. Fortune Magazine, Special Issue, Spring/Summer.

Striving for Manufacturing Excellence. 1990. AT&T Technical Journal 69(4)July/August.

Sullivan, E. 1986. OPTIM: Linking cost, time and quality. Quality Progress April:52-55.

Taguchi, G., and Y. Wu. 1980. Introduction to Off-Line Quality Control. Nagoya: Central Japan Quality Association.

Taguchi, G., E. A. Elsayed, and T. C. Hsiang. 1989. Quality Engineering in Production Systems. New York: McGraw-Hill Book Company.

Taylor, F. W. 1911. Shop Management. New York: Harper Brothers.

Taylor, F. W. 1934. The Principles of Scientific Management. New York: Harper Brothers.

Textile World. 1989. World class manufacturers cut labor needs in half. October:71-76.

Thurow, L. 1992. Head to Head: Coming Economic Battles Between Japan, Europe, and America. New York: Morrow and Co.

Tushman, M. L., and W. L. Moore, eds. 1988. Readings in the Management of Innovation. Cambridge, Mass.: Ballinger Publishing Company.

Tversky, A., and D. Kahneman. 1981. The framing of decisions and psychology of choice. Science 211(30 January):453-458.

Voelcker, H. B., A. A. G. Requicha, and R. W. Conway. 1988. Computer applications in manufacturing. Annual Review of Computer Science 3:349-387.

Vollmann, T. E., Berry, W. L., and D. C. Whybark. 1988. Manufacturing Planning and Control Systems, 2nd ed. Homewood, Ill.: Dow Jones-Irwin.

Voyer, J. J., and R. R. Faulkner. 1989. Organizational cognition in a jazz ensemble. Empirical Study of the Arts 7:57-77.

Walker, A. H., and J. W. Lorsch. 1970. Organizational choice: Product versus function. Pp. 36-53 in Studies in Organization Design. J. W. Lorsch and P. R. Lawrence, eds. Homewood, Ill.: Richard D. Irwin, Inc., and The Dorsey Press.

Walton, R. E. 1969. Interpersonal Peacekeeping: Confrontation in Third Party Consultation. Reading, Mass.: Addison-Wesley.

Whitney, D. E. 1988. Manufacturing by Design. Harvard Business Review July-August.

Whitt, W. 1983. The queueing network analyzer. Bell System Technical Journal 62(9):2779-2815.

Wight, O. 1974. Production and Inventory Management in the Computer Age. Boston, Mass.: Cahners Publishing Co.

Wills, G., and C. Cooper. 1988. Pressure Sensitive. Beverly Hills, Calif.: Sage Publications.

Womack, J.P., D. T. Jones, and D. Roos. 1990. The Machine that Changed the World. New York: Rawson Associates.

Wright, T. P. 1936. Factors affecting the cost of airplanes. Journal of Aeronautical Science 3:122-128.

Committee Membership

Committee on Foundations of Manufacturing

W. DALE COMPTON, Lillian M. Gilbreth Distinguished Professor of Industrial Engineering, Purdue University (Committee Chairman)
H. KENT BOWEN, Ford Professor of Engineering, Massachusetts Institute of Technology
HARRY E. COOK, Grayce Wicall Gauthier Professor, University of Illinois, Urbana-Champaign
JAMES F. LARDNER, Chairman of the Manufacturing Studies Board of the National Research Council
A. ALAN B. PRITSKER, Chairman and Chief Executive Officer, Pritsker Corporation

Contributing Authors

NANCY L. BADORE, Manager, Management and Organization Planning, Employee Relations Staff, Ford Motor Company
HAROLD E. EDMONDSON, Vice President of Manufacturing, Hewlett-Packard Corporation
PHILIP A. FISHER, Fisher & Company
JOHN E. GIBSON, Commonwealth Distinguished Professor of Systems Management, University of Virginia-Charlottesville
WILLIAM C. HANSON, Vice President, Logistics, Digital Equipment Corporation

DAN C. KRUPKA, Department Head, Manufacturing Systems Engineering, AT&T Bell Laboratories
JOHN D. C. LITTLE, Institute Professor, Massachusetts Institute of Technology, Sloan School of Management
DAVID B. MARSING, Plant Manager, Intel Corporation
JOE H. MIZE, Regents Professor of Industrial Engineering and Management, Oklahoma State University
JAMES J. SOLBERG, Director, Engineering Research Center for Intelligent Manufacturing Systems and Professor of Industrial Engineering, Purdue University
G. KEITH TURNBULL, Vice President, Technology Planning, Alcoa
ALBERTUS D. WELLIVER, Corporate Senior Vice President, Engineering and Technology, The Boeing Company
RICHARD WILSON, Professor Emeritus, Industrial Engineering and Operations Department, University of Michigan

NAE STAFF

JOSEPH A. HEIM, Study Director, NAE J. Herbert Hollomon Fellow
H. DALE LANGFORD, Editor
BRUCE R. GUILE, Director, Program Office
MARY JAY BALL, Administrative Assistant

Biographies of Contributing Authors

NANCY L. BADORE is manager of management and organization planning on the Employee Relations Staff, Ford Motor Company. She is responsible for corporate recruiting and placement, planning and monitoring the company's management development system, and overseeing the organization planning of Ford. Dr. Badore was one of four women profiled in Sally Helgeson's book *The Female Advantage: Women's Ways of Leadership.* She received her Ph.D. in social psychology from Boston College.

JOHN R. H. BLACK is a technology planning associate at the Aluminum Company of America. Since joining Alcoa in 1985, he has been involved in planning and plant operations analysis across the gamut of integrated businesses of the company. Before 1985 he did consulting work in Boston in metals and minerals market analysis and forecasting. He received his education at the Massachusetts Institute of Technology, completing a Ph.D. in materials engineering in 1980.

H. KENT BOWEN is Ford Professor of Engineering at the Massachusetts Institute of Technology. As codirector of MIT's Leaders for Manufacturing Program, he guides a research and education program developing the fundamentals for "big-M" manufacturing. His past research included studies of advanced materials and materials processing. He received his B.S. degree in ceramic engineering from the University of Utah and his Ph.D.

degree from MIT. Dr. Bowen is a member of the National Academy of Engineering.

W. DALE COMPTON is the Lillian M. Gilbreth Distinguished Professor of Industrial Engineering at Purdue University. Dr. Compton is a member of the National Academy of Engineering and was the first senior fellow at the NAE. He was vice president for research at the Ford Motor Company. Before joining the Ford Motor Company, Dr. Compton was a professor of physics and director of the Coordinated Science Laboratory at the University of Illinois. He holds degrees in physics from Wabash College (B.A.), the University of Oklahoma (M.S.), and the University of Illinois (Ph.D.).

HARRY E. COOK is the Grayce Wicall Gauthier Professor of Mechanical and Industrial Engineering at the University of Illinois at Urbana-Champaign and director of the Manufacturing Research Center for both the Chicago and the Urbana-Champaign campuses of the university. Dr. Cook, whose research area is lead time reduction and competitiveness theory, has more than 17 years of experience in the automobile industry. He was general manager of scientific affairs at Chrysler Corporation before joining the University of Illinois in June 1990. Dr. Cook is a fellow of the American Society for Metals and a member of the National Academy of Engineering.

ARNOLDO R. CRUZ is manager of technology planning for the Aluminum Company of America. Mr. Cruz joined Alcoa in 1971 as a chemical engineer in Point Comfort, Texas, and has served as production supervisor, production superintendent, and alumina process superintendent while in Point Comfort. Mr. Cruz holds B.S. and M.S. degrees in chemical engineering from Ohio University and an M.S. degree in industrial engineering from Purdue University.

MICHELLE D. DUNLAP is an internal consultant in new products reliability and operations quality for Cummins Engine Company. She is responsible for statistical and reliability analysis for all on- and off-highway diesel engines and engine component groups. She received her B.A. in mathematics from Indiana University and her M.S. in industrial engineering from Purdue University.

HAROLD E. EDMONDSON is vice president of manufacturing for the Hewlett-Packard Company. He has been with HP for 36 years and has held a number of management positions in manufacturing, marketing, and general management. Before taking his current job, he was general manager of the Microwave and Communications Group. Mr. Edmondson is a lecturer in the Sloan Program at the Stanford Graduate School of Business. He received his B.S. degree in mechanical engineering from the University of Kansas and his MBA from Harvard Business School.

EDEN S. FISHER is a technology planning specialist for the Aluminum Company of America. She has been involved in the development and implementation of business analysis, technology planning, and quality management tools at Alcoa since 1984. Before joining Alcoa, she was a postdoctoral fellow at the U.S. Environmental Protection Agency. She holds an A.B. degree in chemistry from Princeton University and a Ph.D. degree in engineering and public policy from Carnegie Mellon University.

PHILIP A. FISHER founded his own business of Fisher & Company in 1931. For the past 45 years, he has focused solely on finding special growth stocks and staying with them as long as they continued to grow substantially more than industry as a whole. Management superiority has become an area of increasing emphasis in these selections. In 1958 he summarized the methods he used in selecting investments together with the investment policies he follows in handling these funds in a book entitled *Common Stocks and Uncommon Profits*. This book has run to some 10 printings in three editions and is believed to be the first book on common stock investing ever to have made the *New York Times* best-seller list. In the 1960s he twice taught the senior class in investments at the Stanford Business School and has frequently been a guest lecturer there since. He graduated from Stanford University and spent one year in their Graduate School of Business.

JOHN E. GIBSON is the Commonwealth Distinguished Professor of Systems Management at the School of Engineering and Applied Science, University of Virginia, Charlottesville. He is past dean of engineering at two universities. His current research is in manufacturing strategy and management and in total quality leadership, and his most recent book is titled *Modern Management of the High Technology Enterprise* (Prentice-Hall, 1990). Dr. Gibson received his Ph.D. from Yale University.

WILLIAM C. HANSON is vice president for logistics at the Digital Equipment Corporation, where his role is to integrate the extended Digital enterprise across the value chains that link engineering, manufacturing, sales, marketing, customers, and suppliers. Before moving to his current position, he was vice president of manufacturing. He is a member of the governing board of directors of the Massachusetts Institute of Technology Leaders for Manufacturing Program and a member of the board of directors of Carnegie Group Incorporated, a leading software and artificial intelligence application organization. He received his bachelor's and master's degrees in industrial engineering from Stanford University.

JOSEPH A. HEIM is the J. Herbert Hollomon Fellow at the National Academy of Engineering. He has worked as a systems engineer for 15 years, concentrating on research, design, and development of hardware and software systems that integrate manufacturing and management functions. He has a bachelor's degree in mechanical engineering, a master of engineering degree in computer science from the University of Louisville, and M.S. and Ph.D. degrees in industrial engineering from Purdue University.

DAN C. KRUPKA has been with AT&T Bell Laboratories since 1967 and is currently head of the Manufacturing Systems Engineering Department. His department works with AT&T's factories to improve their manufacturing operations. Dr. Krupka has a bachelor of engineering degree in engineering physics from McGill University, a Ph.D. in experimental physics from Cornell University, and an advanced professional certificate in economics from New York University.

JAMES F. LARDNER recently retired from Deere & Company as vice president of tractor and component operations. During his 44 years with Deere, he held a wide variety of engineering and manufacturing assignments in both domestic and foreign operations. These involved the design, construction, and start-up of new facilities and the management of factories and design groups in both domestic and overseas divisions. In the last 15 years, he led the corporate effort to reintegrate manufacturing in the Deere organization and to identify and promote the effective use of computer-based tools in design and manufacturing. He has a bachelor of mechanical engineering degree from Cornell University. Mr. Lardner is a member of the National Academy of Engineering and is currently chairman of the National Research Council's Manufacturing Studies Board.

JOHN D. C. LITTLE is Institute Professor and Professor of Management Science at the Massachusetts Institute of Technology. He has done research in queueing theory, mathematical programming, traffic signal synchronization, marketing, and decision support systems. He received an S.B. degree in physics and a Ph.D. in operations research, both from MIT.

DAVID B. MARSING is a plant manager at Intel Corporation. He manages the Albuquerque, New Mexico, facility that produces all of the 80486 and most of the 80386 microprocessors. He has been with Intel for 10 years and has been director of factory automation as well as manager of Intel's facility in Livermore, California. He has a B.S. degree in physics from the University of Oregon, where he also did graduate studies in physics and business.

JOE H. MIZE is Regents Professor of Industrial Engineering and Management at Oklahoma State University, where he is also the director of the Center for Computer Integrated Manufacturing. His major interests are strategic planning for advanced manufacturing systems, design of integrated manufacturing systems, and the design of object-oriented modeling environments for the simulation of complex systems. Dr. Mize is a former president of the Institute of Industrial Engineers. He received his Ph.D. in industrial engineering from Purdue University.

MARYALICE NEWBORN recently started a consulting firm, Corporate Strategies International, Ltd. From 1981 to 1991 she worked for the Aluminum Company of America in areas of fundamental research, advanced manufacturing strategy, technology planning, and corporate strategy. Before that, she worked for Westinghouse as a project manager for naval nuclear contracts. Ms. Newborn holds a B.S. in mechanical engineering from Marquette University, an M.S. in mechanical engineering from Carnegie Mellon University, and an MBA from Duquesne University, and has done graduate work in computer science from National Technological University.

EILEEN M. PERETIC is information strategy director for the Aluminum Company of America, where she is responsible for directing the development of an information policy and implementation plan for Alcoa. Mrs. Peretic joined Alcoa in 1979 and has served as a management information systems analyst, R&D computing analyst, and technology planning specialist. She holds a B.S. degree in industrial engineering from the University of Pittsburgh.

A. ALAN B. PRITSKER is chairman of Pritsker Corporation. He has been actively involved in the development of modeling and simulation languages while employed at Battelle Memorial Institute, Arizona State University, Virginia Polytechnic Institute and State University, and Purdue University. He has published more than 100 technical papers and 9 books on industrial engineering. He received his B.S. degree in electrical engineering and an M.S. degree in industrial engineering from Columbia University, and his Ph.D. degree from the Ohio State University. Dr. Pritsker is a member of the National Academy of Engineering.

JAMES J. SOLBERG is director of the Engineering Research Center for Intelligent Manufacturing Systems and a professor of industrial engineering at Purdue University. His research interests include stochastic processes, mathematical modeling, and manufacturing systems. He received a B.A. degree from Harvard College in mathematics and M.S. and Ph.D. degrees in industrial engineering from the University of Michigan. Dr. Solberg is a member of the National Academy of Engineering.

G. KEITH TURNBULL is vice president of technology planning for the Aluminum Company of America. Dr. Turnbull joined Alcoa in 1962 as a research engineer in the castings and forgings division of the Cleveland Works. His experience in Alcoa includes both technical and management positions at Alcoa's Cleveland Works and at Alcoa Laboratories and Alcoa headquarters in Pittsburgh before being named to his present position in 1986. Dr. Turnbull holds B.S. and M.S. degrees in metallurgical engineering, and has a Ph.D. in physical metallurgy from Case Western Reserve University.

ALBERTUS D. WELLIVER is senior vice president for engineering and technology at The Boeing Company. He has conducted extensive research into all aspects of aircraft propulsion systems and worked on the development of the Boeing 747 propulsion systems installation as well as the CX, SST, supersonic tactical aircraft, and other programs. Before joining Boeing, he worked at the Research Division of Curtiss-Wright Corporation. He received his B.S. degree in mechanical engineering from Pennsylvania State University and completed the Stanford University Executive Business Program. Mr. Welliver is a member of the National Academy of Engineering.

RICHARD C. WILSON is Emeritus Professor in the Department of Industrial and Operations Engineering at the University of Michigan in Ann Arbor. Since retiring from teaching and research in manufacturing planning and facility design, he has been active as a consultant and now a jazz trombonist in the Ann Arbor area.

Index

A

Accreditation Board for Engineering and Technology (ABET), 156–157
Accreditation process, 156–157
Adaptive control, 200
Administrative theory, classical, 117–119
Alcoa, 225, 232
Artificial intelligence
  use of simulation models and, 207
  used for schedule management, 211

B

Badore, Nancy L., 27, 36, 37, 85
Bateson, G., 62
Benchmarking. *See also* Metrics
  competitive, 200
  explanation of, 100, 108
  importance of, 63
  metrics useful for, 100
  process performance and, 229
Blakey, Art, 239
Bloch, E., 16, 73
Boeing, 235–236
Bowen, H. Kent, 54, 72, 93
Breakthrough planning, 200
Bridges, Bill, 192

C

Camp, R. C., 100
Capacity management. *See* Total capacity management (TCM)
Capacity requirements analysis, 209–210. *See also* Total capacity management (TCM)
Capital expenditures, 147
Change
  ongoing, 208, 233

promotion and management of, 233–237
risk as consequence of, 190
Change control system, 194
Chew, W. B., 48
Clark, K. B., 202
Commercialization projects, 94
Communication
and change process, 235
effect of functional groups on, 175
employee motivation through, 236–237
as key element of success, 79
transformation in manufacturing due to technological advances in, 10
Communication barriers, 160
due to different manufacturing languages, 175–176
information management systems and, 176–178
between material sourcing and procurement, 174
between product design and production, 173–174
regarding product support, 174
Communications network
to facilitate distribution of knowledge and information, 74
as principle of integrated enterprise, 161–162
Competition
goals of manufacturers regarding, 28
and time-pressure, 67
understanding capability of, 29
Competitive advantage, 68, 81, 147–148
Competitive benchmarking, 200
Competitive capability, long-term, 104–105
Competitive environment, worldwide, 10–11, 78–79, 85
Compton, W. Dale, 46, 50, 53, 55, 107
Computer-integrated manufacturing (CIM), 204
Computer modeling, 75
Computer technology, 10, 215
Concurrent engineering, 18. *See also* Simultaneous engineering
Conniff, Ray, 242
Continuous improvement, 200
Cook, Harry E., 33, 45, 48, 72–73, 116
Cost performance, 109, 168
Costs
communication problems regarding, 176
learning curve related to, 109–110
Critical variables, 57–58
Cross-functional teams, 172
Customer awards, 143
Customer complaints, 143–144
Customer satisfaction
checks on, 146
elements of, 128
for inside customers, 133
leadership in issues dealing with, 164
meaning of, 130–131
objectives involved in, 131–132
Customers
defining organization's, 128–130
as members of manufacturing system, 79
served by manufacturing organizations, 4–5, 30–31
understanding of, 63
Cycle time
in integrated enterprise, 159
reduction in, 146
variance of, 47

D

Data, strategic analysis, 225–226
Davis, S., 197, 199
Decision-expenditure curves, 53–54
Design
communication barriers between production and product, 173–174
simulation and assessment of, 208–209
Design process, 150–153
Dispatch lists, 206, 207
Double-loop learning, 63
Drucker, P. F., 32, 65
Dunlap, Michelle D., 107

E

Economic lot size model, 185
Economic warfare, 12
Edmondson, Harold E., 31, 51, 63, 128
Education. *See also* Engineering education
establishment of quantitative performance objectives and goals for, 154–157
experienced-based, 178
implications of manufacturing foundations for, 26
insufficiencies in business and engineering, 149–150
need for reassessment of, 80–81
Eisenberg, E., 240
Ellington, Duke, 242
Empirical models
explanation of, 52–54
laws vs., 182–185
Employee capability, 29
Employee empowerment
critical need for, 87, 188
employee involvement vs., 86–87
importance of, 34, 37
in integrated enterprise, 161
Employee involvement
critical nature of, 87
employee empowerment vs., 86–87
environment encouraging, 79–80
explanation of, 86
importance of, 36–37, 147
objectives of, 35–36
at plant level, 88
risk and, 190
Employee motivation
management climate and, 141
through communication, 236–237
Employees
assessment of, 105
as asset of organization, 5, 35
as customers, 30–31
participation in strategic analysis by, 231–232
understanding of manufacturing foundations by, 24
Empowerment. *See also* Employee empowerment
of students, 155
by world-class manufacturers, 35–37, 80
Engineering
coordination between manufacturing and, 144
and design process, 150–153
responsibility involved in, 133
Engineering education
steps to reduce Taylorism in, 154–157
Taylorism as obstacle to, 150–153
Engineering limits, 57
Engineers Council for Professional Development, 156
Entrepreneurial surveillance, 200

Ephlin, Donald F., 61, 89
Experience curve. *See* Learning curve

F

Factories
  design of future, 59
  ongoing modernization of, 196
Feedback loops
  influences on, 54
  learning and, 63–64
15/85 rule, 54, 58, 67, 94, 96, 98
Financial metrics
  explanation of, 48–50
  used for benchmarking, 101
Fisher, Philip A., 29–30, 37, 68, 137
Ford Executive Development Center, 92
Ford Motor Company, 88–92
Forecasts
  development of, 226–229
  summarizing, 230–231
Foreign competition, 10–11
Foster, R., 71
Functional filtering, 134

G

Garvin, D. A., 108
Gemba, 208, 234
General Electric Medical Systems, 47
Gibson, John E., 19, 80
Goals
  to adopt foundations of world-class manufacturing systems, 82
  for education, 154–157
  metrics to define, 51
  short- and long-term, 29
  of world-class manufacturers, 4, 28–30
Goodman, Benny, 243

H

Hall, Jim, 239
Hanson, William C., 28–30, 33, 37, 40, 74, 158
Hayes, R. H., 202
Heim, Joseph A., 107
Herman, Woody, 243
Hewlett–Packard, 46–48
Historical time series graphs, 227
Holland, Dave, 239
House, C. H., 48

I

Imai, Masaaki, 233–234
Improvement
  determining limits for, 55, 57
  organizational ability for, 61
Incentives, 41
Industrial operations engineering, 172
Information-based organizations, 65
Information management systems, communication barriers and, 176–179
Integrated enterprise
  framework for, 159–161
  future of, 164–165
  leadership in, 163–164
  manufacturing as, 158–159
  measures in order to attain, 33
  organizational strategy and, 162–164
  principles of, 161–162
Intel, 193, 194
Involvement. *See* Employee involvement

J

Japanese manufacturing, 32, 134
  change in, 233

product introduction in, 169
quality control in, 144
relationship with supplier and purchaser in, 37
vendor dealings of, 145
Jazz, as metaphor for high-performance teams, 238–244
Johnson, H. T., 49
Just-in-time inventory, 45
Just-in-time manufacturing, 228

K

Kaizen, 208, 233–234
Kaplan, R. S., 49
Kelvin, Lord, 43
Knowledge
corporate memory and, 200
experience-based vs. theory-based, 25
power and, 234–236
Krupka, Dan C., 19, 53, 57–58, 76–77

L

Labor strikes, 145
Lardner, James F., 32–33, 58, 60, 74, 173
Laws. *See also* Manufacturing laws
empirical models vs., 182–185
explanation of, 51–52
physical, 181–182
tautologies vs., 180–182
Leadership
in integrated enterprise, 163–164
knowledge and qualities of successful, 78–79
organization size and, 40
people, 161
product quality and, 41
technology, 161
Lean production, metrics used for, 45
Learning
foundations related to, 62–68
organizational, 61, 199–203
performance improvement as result of, 7
Learning curve
conclusions regarding, 114–115
observations regarding, 111–113, 182–183, 186
related to costs, 109–110
related to quality, 110–111
Learning curve models, 50, 53
Learning organization, strategic control in, 199–203
Learning process, 62–64, 66
Limits
engineering, 57
recognition of technological, 71
theoretical, 57
Linear programming, duality in, 185
Little, John D. C., 52–53, 55, 180, 222
Long-term goals, 29
Long-term investors, 139, 142
Loucks, Vernon R., Jr., 78

M

Magnetic resonance imaging (MRI) systems, 47
Malcolm Baldrige National Quality Award, 39
Management
future vision in, 203
hierarchical and autocratic trends in, 175
impact of short-range investors on, 138–139
implementation of changes in senior, 91–92

in integrated enterprise, 162–164
Japanese leadership in, 32
need for future vision in, 197–198
and process of change, 234–235
responsibility of, 6
strategic planning and, 58
traditional vs. change-driven, 197, 198
understanding tasks of, 38, 40, 42
Management accounting systems, 49
Manne, Shelley, 239
Manufacturers. *See also* World-class manufacturers
characteristics of post-World War II, 85–86
goals and objectives of, 4
measures for rating, 117–118
realistic goals for, 55, 57
Manufacturing
conflicting interests in, 27–28
cooperation between research and, 140–141
coordination between engineering and, 144
future of competitive, 25, 160, 164–165
key terms used in, 176
life cycle and business cycle in, 189
management in future, 65
product definition and, 133
science of, 16
Taylorism as obstacle to, 150–153. *See also* Taylorism
Manufacturing executives
education for, 178–179
questions asked by security analysts of, 142–148
Manufacturing foundations. *See also* World-class manufacturers
audience, 24, 26
benefits of, 23–24, 26
explanation and importance of, 20–23
Manufacturing laws
empirical models vs., 182–185
outlook for, 185–186
tautologies vs., 180–182
Manufacturing opportunity, 230–231
Manufacturing process technologies, 56
Manufacturing processes
complexity of, 215
diagrams of, 168–169
performance forcasting, 227–231
use of models to understand, 208
Manufacturing resource planning (MRP II), 206, 212
Manufacturing systems. *See also* Integrated enterprise
determining optimal strategy for, 43–44
development of theories in, 75
enhancement of scientific method for understanding, 75–77
explanation of, 15–20
identification of laws of, 52
identification of variables of, 57–58
interdependencies of, 7, 15, 17, 79, 149
modeling of, 186–188, 205
study of, 14–15
transformation of U.S., 9–10
Manufacturing unit, 218–220
Marketing responsibilities, 132–133
Marsalis, Wynton, 238–239
Marsing, David B., 36, 40, 46, 72, 189
Mathematical models, simulation used in, 75–76
Mathematical tautologies, 180

Matsushita, Konosuke, 187–188
Metrics
  choice of fundamental, 119
  to define goals and performance expectations, 51
  to define speed of learning, 64
  direct, 45–46
  to evaluate product plan, 135
  evaluation of competitors, 29
  explanation of, 44–45
  financial, 48–50
  interactions between, 106
  as operational guidelines, 50–51
  for professional education, 154–155
  proxy, 45
  sources for, 44
  taxonomy for, 45–46
  use of appropriate, 46–48
Mitchell, Grover, 242
Mize, Joe H., 34–35, 58, 63–64, 196
Model calibration, 200
Modeling
  facilitation of, 205
  growth in use of, 75
  of manufacturing systems, 205
Models
  as basis for decisions and performance prediction, 58–60
  conditional relations in manufacturing system, 75–76
  dangers in use of, 216–217
  empirical, 52–54, 182–185
  explanation of, 52, 204–205
  power of simple, 217–223
  and understanding of critical variables, 57–58
  used for exploration of strategic alternatives, 58
  used to formalize organizational knowledge, 66–68
  user understanding of, 54–55
Motivation, employee, 141, 236–237

N

Nadler, G., 59
National Center for Manufacturing Sciences (NCMS), 11
National Critical Technologies Panel, 69
Nonfinancial metrics, 49–50

O

Objectives
  importance of short-term, 30
  of world-class manufacturers, 4, 28–30
Organization charts, 162, 163
Organization size
  and access to technology, 81–82
  advantages of small, 87
  leadership and, 40
Organizational behavior, 160
Organizational culture, 116
Organizational development model, 89
Organizational knowledge, 66–68
Organizational structure
  criteria relevant to, 117–118
  as organizational barrier, 160
Organizations
  dissatisfaction with functional, 116, 117
  employees as asset of, 5
  impact of, 31–35
  information-based, 65
  renewal of, 62
  transition process in, 192–193

P

Participative management
  of education enterprise, 156
  high-performance teams in, 238–244

implementation of, 90–91
People leadership, 161. *See also* Leadership
Performance
aggregated measures of, 104
metric of time, 171–172
Performance evaluation
explanation of, 44
process of, 6
Pestillo, Peter, 89
Physical laws, 181–182, 205
Power, knowledge and, 234–236
Price, R. L., 48
Pritsker, A. Alan B., 59, 66–67, 204
Problem-solving methods, 193
Process control systems, 193
Process improvement programs, 221
Process introduction
empirical observations for, 94–99
requirements for, 93
Process performance, forecasting, 227–229
Product definition, 132–133
Product introduction
empirical observations for, 94–99
requirements for, 93
time as element in, 169, 172
Product performance metrics, 101–102
Product plan
checks regarding, 1136
customer satisfaction and, 132–135
implementation of, 136
Product quality, 41
Product unit, 218
Production
learning curves and, 111
quality index and cumulative volume of, 114
use of simulation for control over, 206
Production process
transfer of responsibility to operators of equipment for, 193–194
use of statistical metrics in, 46
Profitability
nonfinancial indicators and long-term, 49
as proxy metric, 45
Prospect theory, 184–185
Proxy metrics, 45

Q

Quality
assessment of trends in, 108
definitions of, 108
importance of, 107–108
learning curve related to, 110–111
measures of, 108–109
Quality circles, 88
Quality control personnel, 144–145
Quality function deployment (QFD) approach, 127, 132
Quality index (QI), 111–114
Queueing models, 169–172
Queueing systems, 180–181
Queueing theory, 52
Quinn, James Brian, 27, 41

R

Research and development
cooperation between manufacturing and, 140–141
dominant role in U.S. of, 134
Research community, study of manufacturing foundations by, 26
Response flow checklists (RFCs), 193–194
Rewards, 41

Risk
ability to handle, 195
avoidance of, 189
as consequence of change, 190
control methodology and, 194
and effective use of people, 190–191
and planning for change, 191–193
statistics, problem solving and, 193–194
Robinson, G. H., 59

S

S–curve phenomena, 228
Satisfaction, 130–131
Scheduling
execution and dispatching in, 211
management of, 210–211
use of simulation for, 206, 210, 211
Schneiderman, A. M., 111
Scientific management principles, 22
Scientific method, 75–77
Securities and Exchange Commission, 101, 139
SEMATECH, 11
Short-range investors, 138–139
Short-term goals, 29
Simon, Herbert A., 74–75, 205
Simulation
large-scale, 186
of manufacturing systems, 60, 205
in mathematical models, 75–76
as mechanism for explaining and distributing complex rules and policies, 59
for scheduling purposes, 206, 210, 211
use of, 60, 75, 206
Simulation languages
availability of, 75
to build and analyze manufacturing models, 205
Simulation modeling, 60
Simultaneous engineering, 18
requirements of, 72
as team concept, 127
Single-loop learning, 62, 63
Size. *See* Organization size
Small manufacturers, 81–82
Solberg, James J., 25, 55, 75, 215
Stacy, Jess, 243
Statistical process control (SPC)
charts used for, 193–194
explanation of, 108, 109
Statistical quality control
explanation of, 143
success of, 142
Statistics
need for working knowledge of, 193
status presentation and, 211
Stock market investors
importance of manufacturing division to, 139–141
types of, 137–139
Strategic analysis, 224
at Alcoa, 225
data in, 225–226
and development of forecasts, 226–229
participation and, 231–232
summarizing forecasts and interpreting opportunities in, 230–231
Strategic control, 199–202
Strategic planning
to create planned crisis, 191–192
early process of, 199, 200
management and, 58
Student empowerment, 155

Student work practices, 155–156
Subsystem interfaces, 72–74
Subsystem research, 70, 72
Subsystem unit functioning, 122–124
Suppliers
barriers between purchasers and, 5–6
relationship with, 37–38
System dynamics, 186–187
System operational metrics, 103–104
System research, 74–75
System/subsystem structure
difference between function and, 124–125
functioning of, 122–124

T

Taguchi's paradigm
application of, 120–124
explanation of, 119–120
Tautologies, 180–182
Taylor, Frederick Winslow, 22, 117, 150–151
Taylorism
appropriateness of application of, 153–154
and education, 154–157
elements of, 150–153
explanation of, 150
Teams
creativity in, 240
large high-performance, 241–244
product development and cross-functional, 172
responsibility needed by, 191
small high-performance, 239–244
trust as element of, 159–160
use of statistics and problem-solving methods by, 193
Technical work force capabilities, 50–51
Technology
acceptance of changes in, 25
assessing developments in, 105
foundations related to, 68, 70–77
Japanese, 32
as key to competitive advantage, 68, 81
leadership in, 161
national critical, 69
recognition of limits to, 71
use of, 7–8, 61, 77
Theoretical limits, 57
Time
as critical metric, 76–77, 166–168
as detector of inefficiencies, 171–172
as diagnostic tool and driver of quality and cost, 168–169
and queueing models, 169–172
Total capacity management (TCM)
architecture of, 212–213
as concept, 214
explanation of, 204
functions in, 208–211
overview of, 206–208
schedule execution and dispatching and, 211
Total quality control (TQC), 235
Transport unit, 219, 220
Transportation technology, 10
Turnbull, G. Keith, 49, 55, 57, 62, 64, 70, 224

U

Union membership, 145
Unit operation metrics, 102–103
United States
ability to take emotional risks and a competitive advantage for, 195
goals to adopt foundations of

world-class manufacturing systems, 82
manufacturing gap in, 73

V

Validity-check model, 55
Vendors
barriers between purchasers and, 5–6
number of and relationships with, 145–146
relationship with, 37–38
Visioning
ability of managers to apply, 197
explanation of, 200
Visualization graphics methods, 75

W

Wall Street investors
importance of manufacturing division to, 139–141
types of, 137–139
Welliver, Albertus D., 36, 38, 63, 64–65, 233
Wheelwright, S. C., 202
Wilson, Richard C., 36, 40, 80, 238
Work force capabilities, 50–51
Work teams. *See* Teams
World-class manufacturers
common model as basis for achievement of, 208
elimination of barriers within organizations by, 35
employee involvement and empowerment by, 35–37, 80
goals and objectives of, 4, 28–30
leadership provided by, 164
meaning of, 28–29
and method of acquiring knowledge, 67–68
models as tools used by, 60
and relationship with customers, 4, 28–33
relationship with suppliers and vendors, 37–38
role of management for, 42
use of metrics to help define goals and performance expectations, 51
view and use of technology by, 77